World Flags

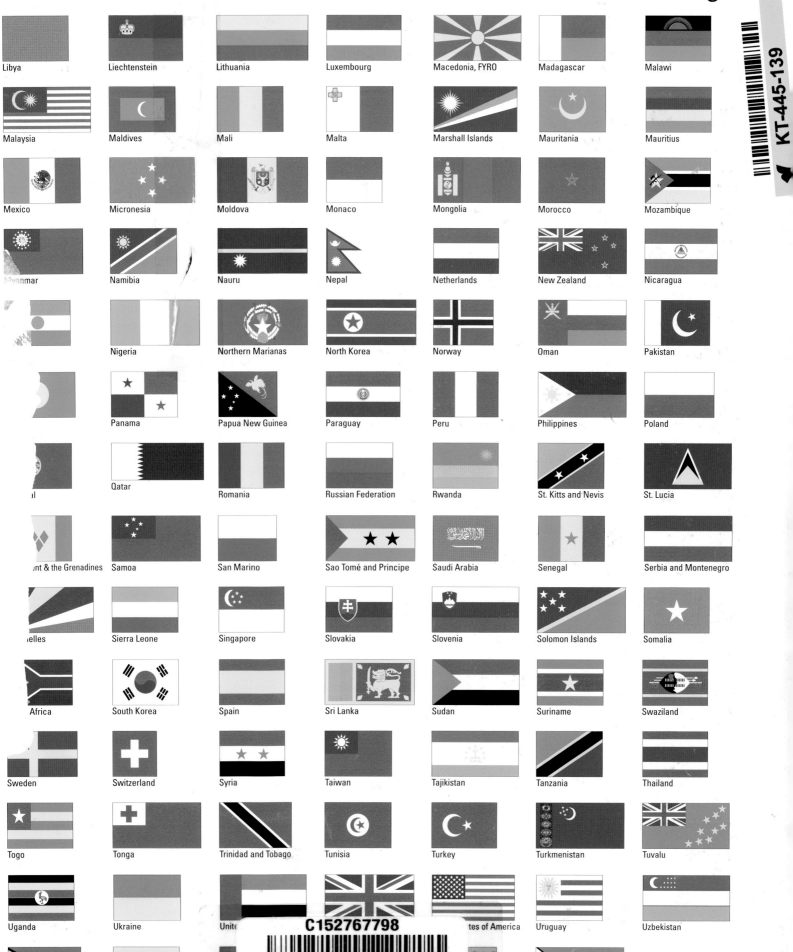

Libya	Liechtenstein	Lithuania	Luxembourg	Macedonia, FYRO	Madagascar	Malawi
Malaysia	Maldives	Mali	Malta	Marshall Islands	Mauritania	Mauritius
Mexico	Micronesia	Moldova	Monaco	Mongolia	Morocco	Mozambique
Myanmar	Namibia	Nauru	Nepal	Netherlands	New Zealand	Nicaragua
	Nigeria	Northern Marianas	North Korea	Norway	Oman	Pakistan
	Panama	Papua New Guinea	Paraguay	Peru	Philippines	Poland
Qatar		Romania	Russian Federation	Rwanda	St. Kitts and Nevis	St. Lucia
ent & the Grenadines	Samoa	San Marino	Sao Tomé and Principe	Saudi Arabia	Senegal	Serbia and Montenegro
elles	Sierra Leone	Singapore	Slovakia	Slovenia	Solomon Islands	Somalia
Africa	South Korea	Spain	Sri Lanka	Sudan	Suriname	Swaziland
Sweden	Switzerland	Syria	Taiwan	Tajikistan	Tanzania	Thailand
Togo	Tonga	Trinidad and Tobago	Tunisia	Turkey	Turkmenistan	Tuvalu
Uganda	Ukraine	Unit[ed]		[Sta]tes of America	Uruguay	Uzbekistan
Vanuatu	Venezuela	Vietnam	Yemen	Zambia	Zimbabwe	

OXFORD
Practical
ATLAS

Editorial Adviser
Dr Patrick Wiegand

OXFORD
UNIVERSITY PRESS

Great Clarendon Street, Oxford OX2 6DP

Oxford University Press is a department of the University of Oxford.
It furthers the University's objective of excellence in research, scholarship,
and education by publishing worldwide in

Oxford   New York

Auckland   Cape Town   Dar es Salaam   Hong Kong   Karachi
Kuala Lumpur   Madrid   Melbourne   Mexico City   Nairobi
New Delhi   Shanghai   Taipei   Toronto

With offices in

Argentina   Austria   Brazil   Chile   Czech Republic   France   Greece
Guatemala   Hungary   Italy   Japan   Poland   Portugal   Singapore
South Korea   Switzerland   Thailand   Turkey   Ukraine   Vietnam

Oxford is a registered trade mark of Oxford University Press
in the UK and in certain other countries

ISBN 0 19 832162 7 (hardback)

ISBN 0 19 832161 9 (paperback)

1 3 5 7 9 10 8 6 4 2

Printed in Singapore

Acknowledgements

The publishers would like to thank the Telegraph Colour Library for permission to reproduce the photograph on page 8.

Cover image: Tom Van Sant / Geosphere Project, Santa Monica, Science Photo Library.

The illustrations are by Chapman Bounford, Hard Lines, and Gary Hinks.

The page design is by Adrian Smith.

2 **Contents** The World, The British Isles

The World

Countries	4–5
Land and Water	6–7
Climate and Ecosystems	8–9
Earthquakes and Volcanoes	10–11
Population	12–13
Development and Interdependence	14–15
Environment and Pollution	16–17

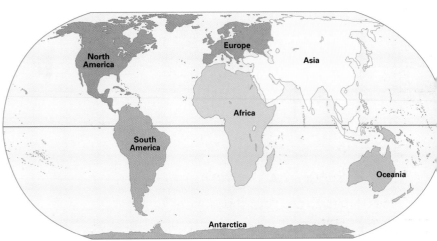

The British Isles

Countries	18
Land and Water	19
Climate	20–21
Population	22
Farming and Industry	23
Energy	24
Communications	25
Pollution	26
Scenery and Conservation	27
Topographic map	28
Orkney and Shetland Islands	29
Scottish Highlands and Islands	30–31
Northern Ireland	32
Southern Scotland	33
Northern England	34–35
Wales	36
English West Midlands	37
South East England	38–39
South West England	40–41

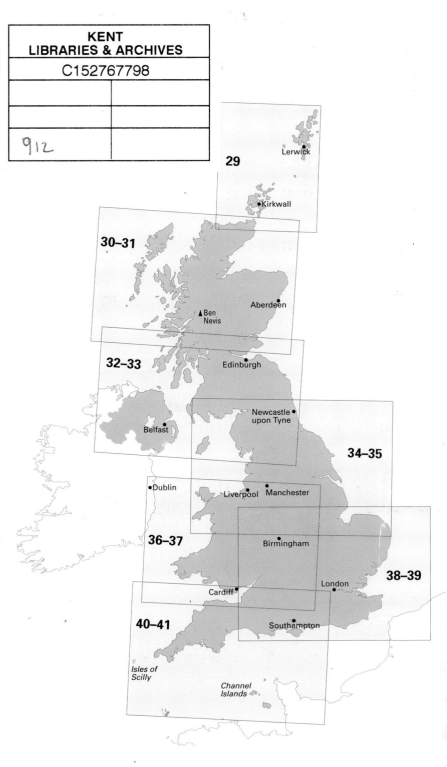

Maps that show general features of
regions, countries or continents are
called **topographic maps.**
These maps are shown with a light
band of colour in the contents list.

For example:

South West England

Contents Continents and Poles 3

Europe

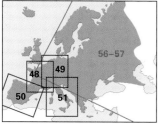

Topographic map	42–43
European Union	43
Wealth	43
Climate and Ecosystems	44
Farming, Energy, and Minerals	45
Industry and Environmental issues	46
Population	47
France	48
Benelux, Germany, Alps	49
Spain and Portugal	50
Italy	51

Asia

Countries, Land and Water	52
Climate and Ecosystems	53
Farming, Energy, Minerals, and Industry	54
Population and Environmental issues	55
Northern Eurasia	56–57
Southern Asia	58
China and Japan	59
South East Asia	60
The Middle East and North Africa	61

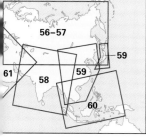

Africa

Countries, Land and Water	62
Climate and Ecosystems	63
Farming, Energy, Minerals, and Industry	64
Population and Environmental issues	65
North and West Africa	66
South and East Africa	67

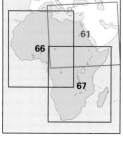

Oceania

Countries, Land and Water	68
Climate and Ecosystems	69
Farming, Energy, Minerals, and Industry	70
Population and Environmental issues	71
Australia	72
New Zealand	73

North America

Countries, Land and Water	74
Climate and Ecosystems	75
Farming, Energy, Minerals, and Industry	76
Population and Environmental issues	77
Canada, Greenland, USA, Mexico and the Caribbean	78–79

South America

Countries, Land and Water	80
Climate and Ecosystems	81
Farming, Energy, Minerals, and Industry	82
Population and Environmental issues	83
South America	84–85

The Poles

The Arctic Ocean	86
Antarctica	86

Index 87-96

Key

CANADA	Country name
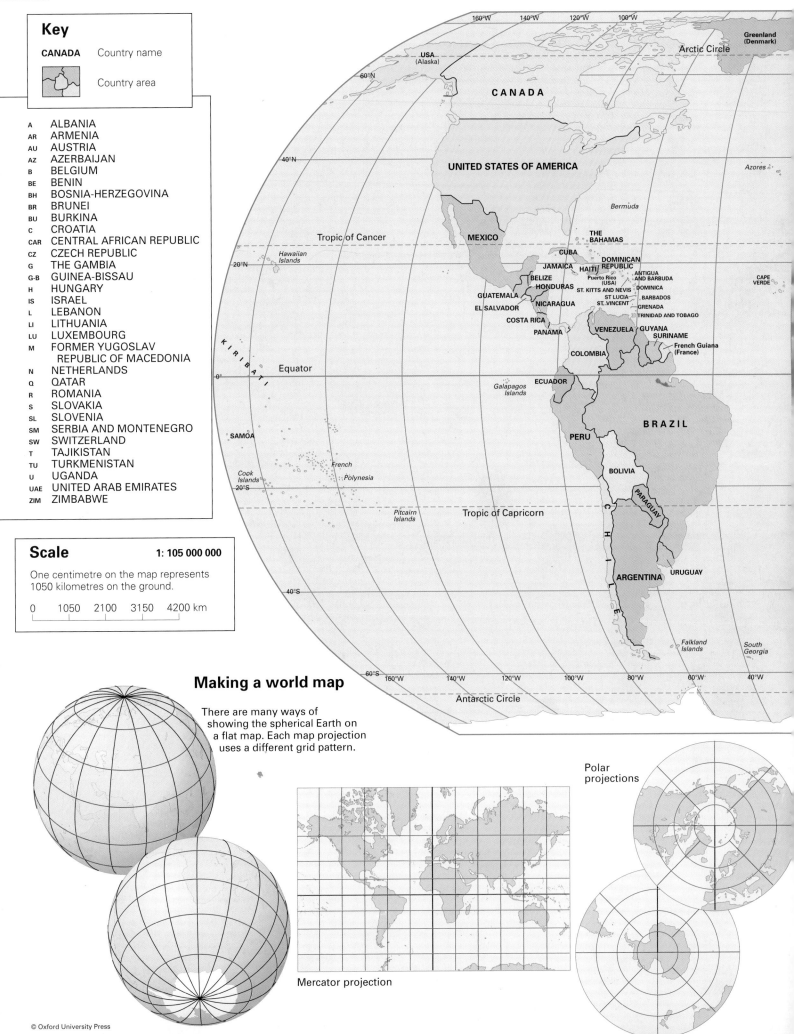	Country area

A	ALBANIA
AR	ARMENIA
AU	AUSTRIA
AZ	AZERBAIJAN
B	BELGIUM
BE	BENIN
BH	BOSNIA-HERZEGOVINA
BR	BRUNEI
BU	BURKINA
C	CROATIA
CAR	CENTRAL AFRICAN REPUBLIC
CZ	CZECH REPUBLIC
G	THE GAMBIA
G-B	GUINEA-BISSAU
H	HUNGARY
IS	ISRAEL
L	LEBANON
LI	LITHUANIA
LU	LUXEMBOURG
M	FORMER YUGOSLAV REPUBLIC OF MACEDONIA
N	NETHERLANDS
Q	QATAR
R	ROMANIA
S	SLOVAKIA
SL	SLOVENIA
SM	SERBIA AND MONTENEGRO
SW	SWITZERLAND
T	TAJIKISTAN
TU	TURKMENISTAN
U	UGANDA
UAE	UNITED ARAB EMIRATES
ZIM	ZIMBABWE

Scale

1: 105 000 000

One centimetre on the map represents 1050 kilometres on the ground.

0	1050	2100	3150	4200 km

Making a world map

There are many ways of showing the spherical Earth on a flat map. Each map projection uses a different grid pattern.

Mercator projection

Polar projections

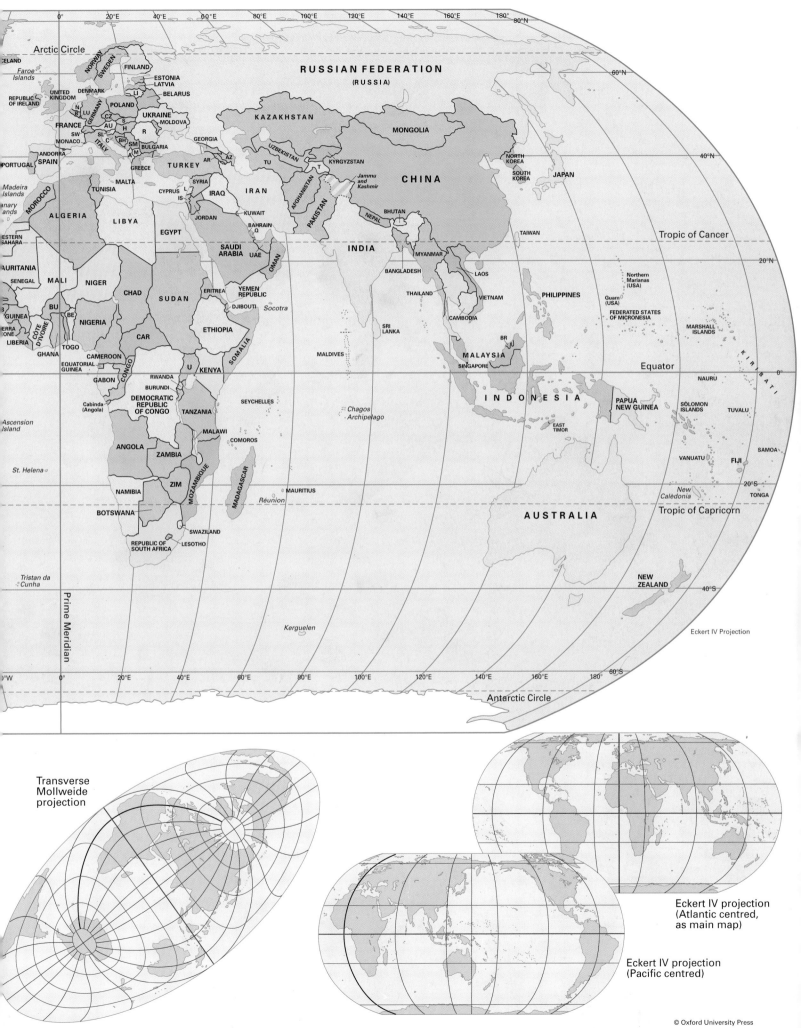

Arctic Circle

ICELAND

Faroe Islands

NORWAY SWEDEN FINLAND

ESTONIA
LATVIA

UNITED
KINGDOM DENMARK LI BELARUS

REPUBLIC
OF IRELAND

N
B LU GERMANY POLAND
CZ UKRAINE

FRANCE SW S R MOLDOVA

SL C H
MONACO AU GEORGIA
ITALY BH SM BULGARIA

ANDORRA A M GREECE

PORTUGAL SPAIN

RUSSIAN FEDERATION
(RUSSIA)

KAZAKHSTAN

MONGOLIA

UZBEKISTAN

TURKEY AR AZ TU T KYRGYZSTAN

SYRIA L CYPRUS IS IRAQ IRAN AFGHANISTAN Jammu and Kashmir

CHINA

NORTH
KOREA

SOUTH
KOREA JAPAN

Madeira Islands

Canary Islands

MOROCCO

ALGERIA LIBYA EGYPT

MALTA

TUNISIA

JORDAN KUWAIT

SAUDI
ARABIA

PAKISTAN

NEPAL BHUTAN

TAIWAN

Tropic of Cancer

40°N

20°N

MAURITANIA

SENEGAL MALI NIGER CHAD SUDAN

BAHRAIN
Q
UAE OMAN

YEMEN
REPUBLIC

INDIA

BANGLADESH

MYANMAR

LAOS

*Northern
Marianas
(USA)*

GUINEA BU BE NIGERIA ERITREA

DJIBOUTI *Socotra*

THAILAND

VIETNAM

*Guam
(USA)*

FEDERATED STATES
OF MICRONESIA

MARSHALL
ISLANDS

SIERRA
LEONE COTE
D'IVOIRE CAR

LIBERIA GHANA TOGO CAMEROON ETHIOPIA

SRI
LANKA

CAMBODIA

EQUATORIAL
GUINEA U KENYA

SOMALIA

MALDIVES

BR

MALAYSIA

Equator

PHILIPPINES

GABON CONGO RWANDA

BURUNDI

*Cabinda
(Angola)* DEMOCRATIC
REPUBLIC
OF CONGO TANZANIA

SINGAPORE

INDONESIA

PAPUA
NEW GUINEA

NAURU

KIRIBATI

SEYCHELLES *Chagos
Archipelago*

SOLOMON
ISLANDS

TUVALU

*Ascension
Island*

MALAWI

COMOROS

EAST
TIMOR

ANGOLA ZAMBIA

St. Helena MADAGASCAR

NAMIBIA ZIM MOZAMBIQUE

Réunion *MAURITIUS*

VANUATU FIJI

SAMOA

*New
Caledonia* TONGA

20°S

BOTSWANA

SWAZILAND

AUSTRALIA

Tropic of Capricorn

REPUBLIC OF
SOUTH AFRICA LESOTHO

*Tristan da
Cunha*

NEW
ZEALAND 40°S

Prime Meridian

Kerguelen

Eckert IV Projection

60°S

Antarctic Circle

Transverse
Mollweide
projection

Eckert IV projection
(Atlantic centred,
as main map)

Eckert IV projection
(Pacific centred)

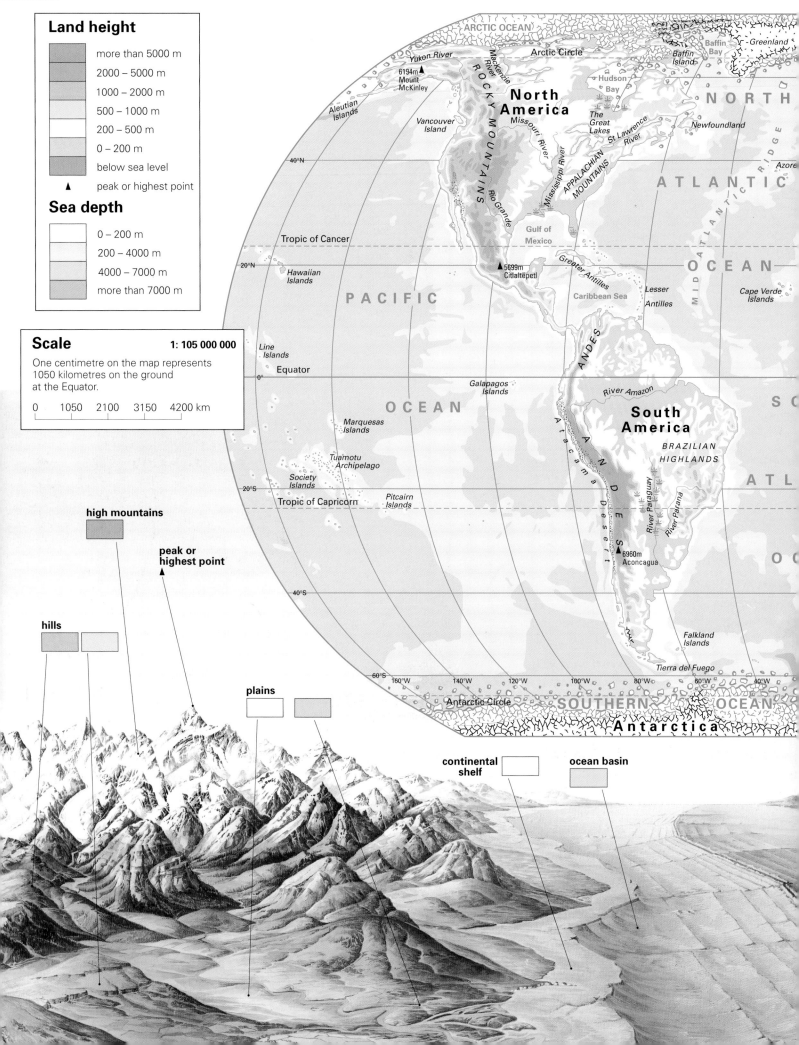

Land height

more than 5000 m
2000 – 5000 m
1000 – 2000 m
500 – 1000 m
200 – 500 m
0 – 200 m
below sea level
▲ peak or highest point

Sea depth

0 – 200 m
200 – 4000 m
4000 – 7000 m
more than 7000 m

Scale
1: 105 000 000

One centimetre on the map represents
1050 kilometres on the ground
at the Equator.

0   1050   2100   3150   4200 km

high mountains

peak or
highest point

hills

plains

continental
shelf

ocean basin

ARCTIC OCEAN
Arctic Circle
Greenland
Baffin
Bay
Baffin
Island

Yukon River
Mackenzie River
ROCKY MOUNTAINS
Hudson
Bay
**North
America**
NORTH

6194m ▲
Mount
McKinley

Vancouver
Island
Missouri River
The
Great
Lakes
St Lawrence
River
Newfoundland

40°N
Mississippi River
APPALACHIAN MOUNTAINS
ATLANTIC

Rio Grande
MID ATLANTIC RIDGE
Azore

Tropic of Cancer
Gulf of
Mexico
OCEAN

20°N
Hawaiian
Islands
5699m ▲
Citlaltépetl
Greater Antilles
Cape Verde
Islands

PACIFIC
Caribbean Sea
Lesser
Antilles

Line
Islands
Galapagos
Islands
River Amazon

0°   Equator

OCEAN
**South
America**
S

Marquesas
Islands
Atacama Desert
ANDES
BRAZILIAN
HIGHLANDS
ATL

Tuamotu
Archipelago

Society
Islands
20°S
River Paraguay
River Paraná

Tropic of Capricorn
Pitcairn
Islands
O

6960m ▲
Aconcagua

40°S
Falkland
Islands

60°S   160°W   140°W   120°W   100°W   80°W   60°W   40°W
Tierra del Fuego

Antarctic Circle
SOUTHERN   OCEAN
Antarctica

ARCTIC OCEAN

Arctic Circle

Eckert IV Projection
© Oxford University Press

Barents
Sea

URAL MOUNTAINS

River Ob

Yenisey River

River Lena

60°N

Sea of
Okhotsk

Bering Sea

Aleutian Trench

North
Sea

Europe

River Volga

River Irtysh

ALTAI MOUNTAINS

Asia

Lake
Baykal

Kuril Trench

British
Isles

R. Rhine

River Danube

Pripet
Marshes

Mount
Elbrus
5642m

Aral
Sea

Gobi Desert

Honshu

40°N

4807m
Mont
Blanc

ALPS

Communism
Peak
7495m

Huang-He

Madeira
Islands

Black Sea

CAUCASUS

Caspian
Sea

8611m
K2

TIBETAN
PLATEAU

Chang Jiang

East
China
Sea

Ryukyu Trench

TAURUS
MOUNTAINS

M e d i t e r r a n e a n   S e a

ZAGROS MOUNTAINS

HIMALAYA

8848m
Mount Everest

Tropic of Cancer

Canary
lands

ATLAS MOUNTAINS

River Nile

Red Sea

River Ganges

Mekong River

South
China
Sea

PACIFIC

20°N

DECCAN

S a h a r a   D e s e r t

Arabian
Sea

Bay of
Bengal

Philippines

Marianas
Islands

Marianas Trench

OCEAN

River Niger

Lake Chad

Andaman
Islands

Yap
Islands

Caroline Islands

Marshall
Islands

Africa

Nicobar
Islands

Philippine Trench

River Congo

Maldive
Archipelago

Sumatra

Borneo

Equator

Gilbert
Islands

0°

Lake
Victoria

5895m
Mount
Kilimanjaro

I N D I A N

Seychelles

New
Guinea

4508m
Mount
Wilhelm

Phoenix
Islands

Lake
Tanganyika

Aldabra
Islands

Java

Solomon
Islands

Samoa
Islands

Lake
Nyasa
(Malawi)

Comoro
Archipelago

O C E A N

Oceania

Espíritu
Santo

Fiji
Islands

M I D - A T L A N T I C

River Zambezi

Madagascar

Great Sandy
Desert

New
Caledonia

Tonga
Islands

Namib Desert

Okavango
Swamp

Mauritius

Réunion

Tropic of Capricorn

Kalahari
Desert

Great Victoria
Desert

GREAT DIVIDING RANGE

Tonga Trench

NULLARBOR PLAIN

River Darling

North
Island

R I D G E

Prime Meridian

Kerguelen
Islands

Murray R.

Tasman
Sea

40°S

Tasmania

3764m
Mount
Cook

South
Island

S O U T H E R N   O C E A N

0°W

0°

20°E

40°E

60°E

80°E

100°E

120°E

140°E

160°E

180°

A n t a r c t i c a

high plateau

low plateau

ocean ridge

continental shelf

**ocean
trench**

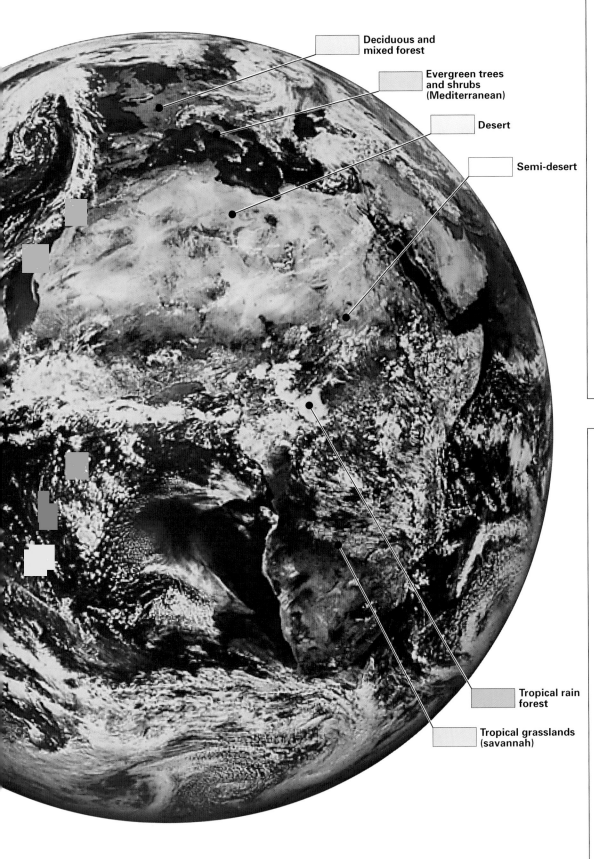

Deciduous and
mixed forest

Evergreen trees
and shrubs
(Mediterranean)

Desert

Semi-desert

Tropical rain
forest

Tropical grasslands
(savannah)

A Meteosat view of
the Earth recorded
by a geostationary satellite
positioned 36 000 km above
the intersection of the
Prime Meridian and the Equator

Climatic regions

Hot tropical rainy

rain all year

monsoon

dry in winter

Very dry

with no reliable rain

with a little rain

**Influenced by the sea:
warm summers, mild winters**

with dry summers
(Mediterranean type)

with dry winters

with no dry season

Cool

with dry winters

rain all year

Cold polar

no warm season
and fairly dry

Mountain

height of the land strongly
affects the climate

Ecosystems

Vegetation types are those which
would occur naturally without
interference by people

Coniferous forest

cone bearing trees

Deciduous and mixed forest

leaf shedding and
coniferous tress

Tropical rain forest

many species of lush,
tall trees

Tropical grasslands (savannah)

tall grass parkland
with scattered trees

Thorn forest

low trees and shrubs with
spines or thorns

Evergreen trees and shrubs

plants and small trees
with leathery leaves

Temperate grasslands

prairies, steppes,
pampas and veld

Semi-desert

short grasses and
drought-resistant scrub

Desert

sand and stones,
very little vegetation

Tundra

moss and lichen,
with few trees

Ice

no vegetation

Mountains

thin soils, steep slopes
and high altitude affects
type of vegetation

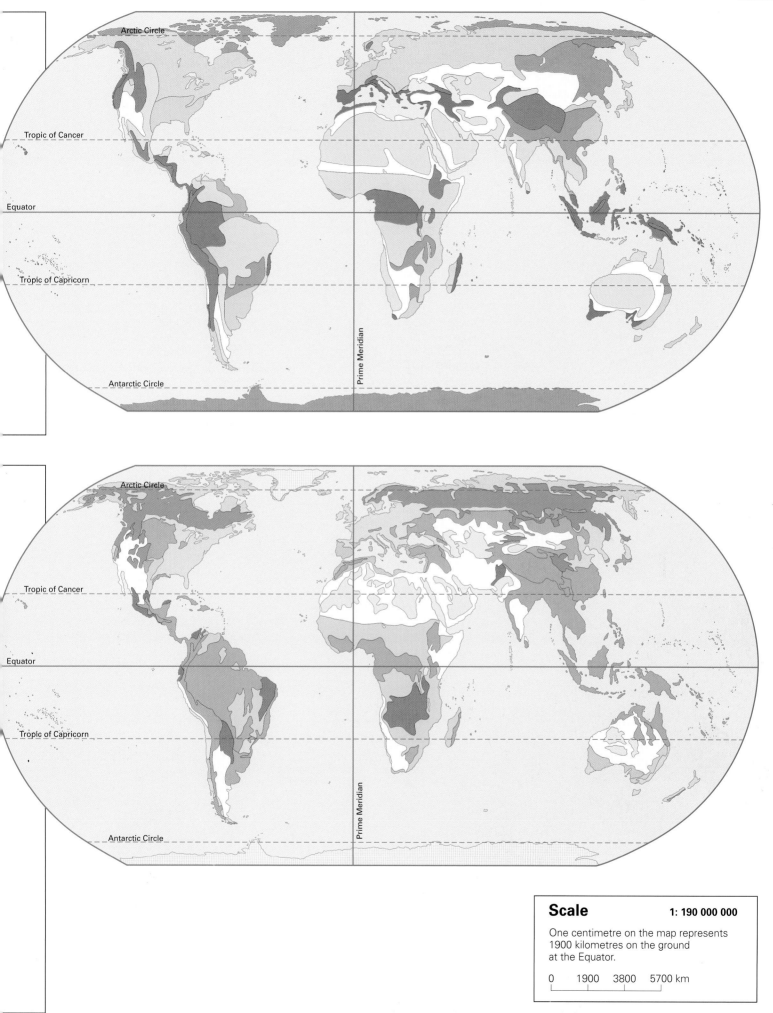

© Oxford University Press

Scale 1: 190 000 000

One centimetre on the map represents
1900 kilometres on the ground
at the Equator.

0   1900   3800   5700 km

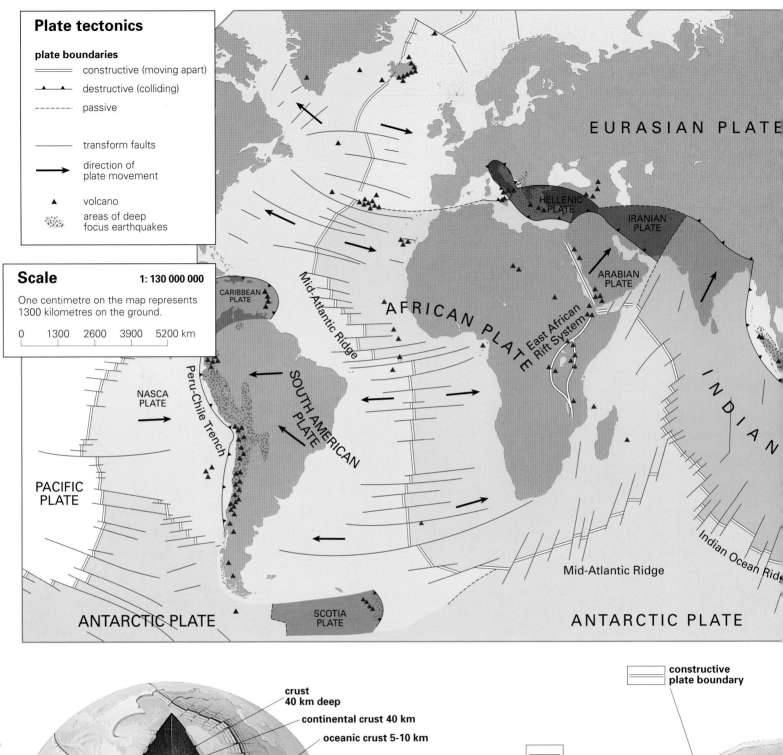

Plate tectonics

plate boundaries

constructive (moving apart)

destructive (colliding)

passive

transform faults

direction of plate movement

▲ volcano

areas of deep focus earthquakes

Scale

1: 130 000 000

One centimetre on the map represents 1300 kilometres on the ground.

0   1300   2600   3900   5200 km

EURASIAN PLATE

HELLENIC PLATE

IRANIAN PLATE

ARABIAN PLATE

CARIBBEAN PLATE

Mid-Atlantic Ridge

AFRICAN PLATE

East African Rift System

INDIAN

NASCA PLATE

Peru-Chile Trench

SOUTH AMERICAN PLATE

PACIFIC PLATE

Mid-Atlantic Ridge

Indian Ocean Ridge

ANTARCTIC PLATE

SCOTIA PLATE

ANTARCTIC PLATE

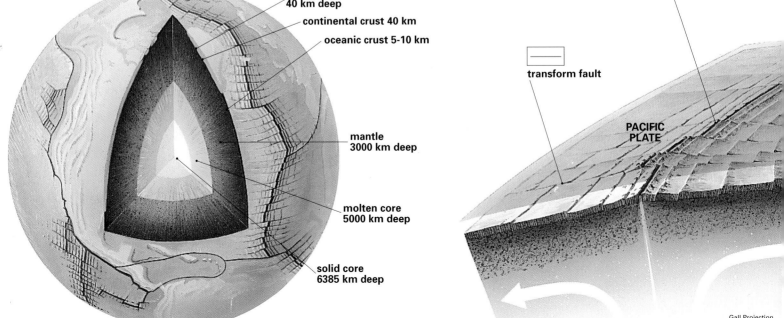

crust 40 km deep

continental crust 40 km

oceanic crust 5-10 km

mantle 3000 km deep

molten core 5000 km deep

solid core 6385 km deep

constructive plate boundary

transform fault

PACIFIC PLATE

Gall Projection
© Oxford University Press

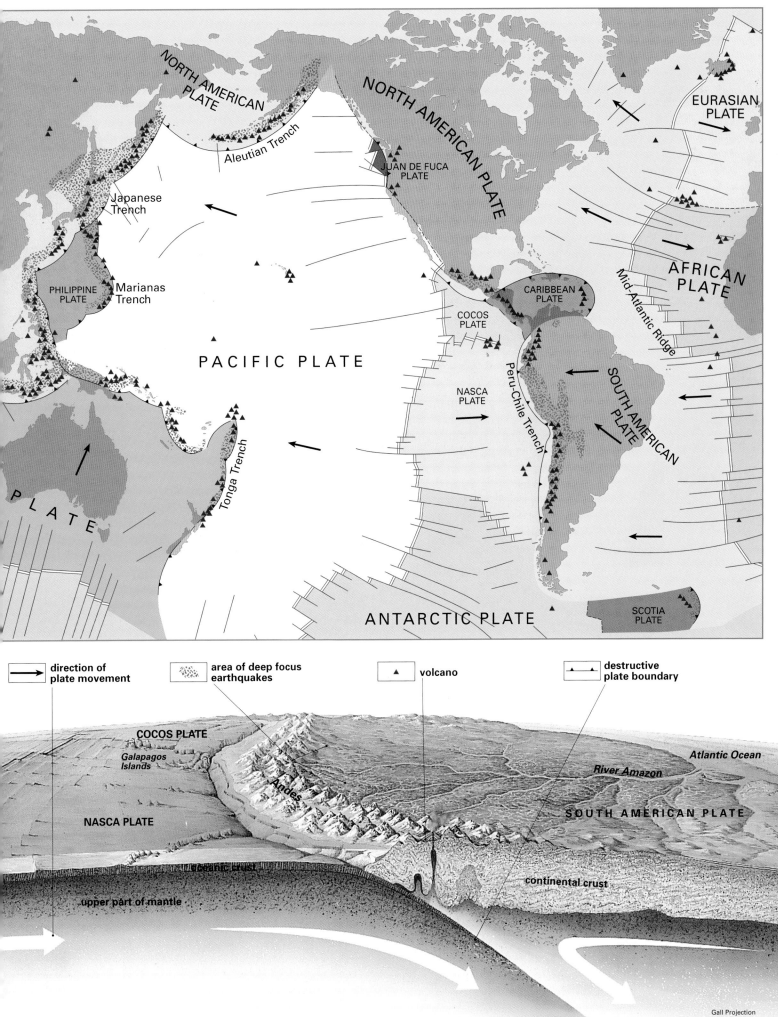

NORTH AMERICAN PLATE

NORTH AMERICAN PLATE

EURASIAN PLATE

Aleutian Trench

JUAN DE FUCA PLATE

Japanese Trench

AFRICAN PLATE

PHILIPPINE PLATE

Marianas Trench

CARIBBEAN PLATE

Mid-Atlantic Ridge

COCOS PLATE

PACIFIC PLATE

NASCA PLATE

SOUTH AMERICAN PLATE

Peru-Chile Trench

PLATE

Tonga Trench

ANTARCTIC PLATE

SCOTIA PLATE

| | direction of plate movement | | area of deep focus earthquakes | | volcano | | destructive plate boundary |

COCOS PLATE

Galapagos Islands

Atlantic Ocean

River Amazon

Andes

SOUTH AMERICAN PLATE

NASCA PLATE

oceanic crust

continental crust

upper part of mantle

Population density

number of people
per square kilometre

high		more than 50
moderate		6 – 49
sparse		1 – 5
very low		less than 1

O major cities and built up
   areas of at least 3 000 000
   people

—— international boundary

Scale   1: 105 000 000

One centimetre on the map represents
1050 kilometres on the ground
at the Equator.

0   1050   2100   3150   4200 km

Population structure of the World

Age
Males   Females

80
70
60
50
40
30
20
10
0

6 5 4 3 2 1 0   0 1 2 3 4 5 6

percent of the population in 2004

In 2004 the total world population
was approximately 6 372 493 257.

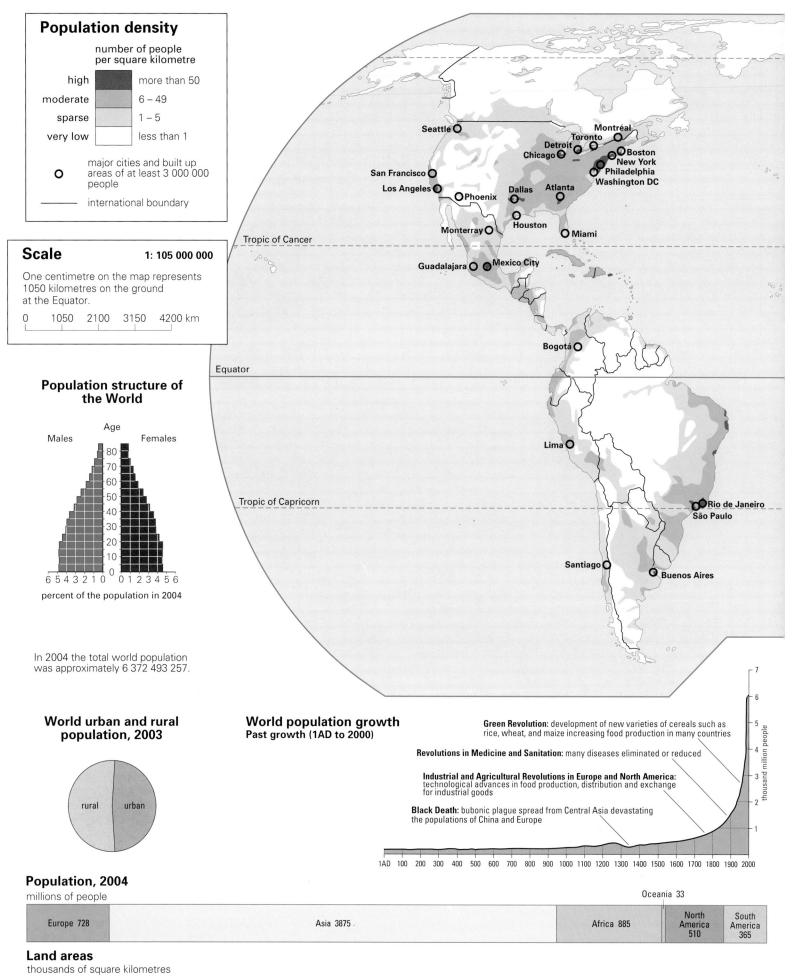

Seattle
Montréal
Toronto
Detroit
Chicago
Boston
New York
Philadelphia
Washington DC
San Francisco
Los Angeles
Phoenix
Dallas
Atlanta
Monterray
Houston
Miami

Tropic of Cancer

Guadalajara
Mexico City

Bogotá

Equator

Lima

Tropic of Capricorn

Rio de Janeiro
São Paulo

Santiago
Buenos Aires

World urban and rural population, 2003

rural   urban

World population growth
Past growth (1AD to 2000)

Green Revolution: development of new varieties of cereals such as
rice, wheat, and maize increasing food production in many countries

Revolutions in Medicine and Sanitation: many diseases eliminated or reduced

Industrial and Agricultural Revolutions in Europe and North America:
technological advances in food production, distribution and exchange
for industrial goods

Black Death: bubonic plague spread from Central Asia devastating
the populations of China and Europe

7
6
5
4
3
2
1

thousand million people

1AD 100 200 300 400 500 600 700 800 900 1000 1100 1200 1300 1400 1500 1600 1700 1800 1900 2000

Population, 2004
millions of people

Oceania 33

Europe 728	Asia 3875	Africa 885	North America 510	South America 365

Land areas
thousands of square kilometres

Europe 10 498	Asia 44 387	Africa 30 335	Oceania 8503	North America 24 241	South America 17 832	Antarctica 13 340

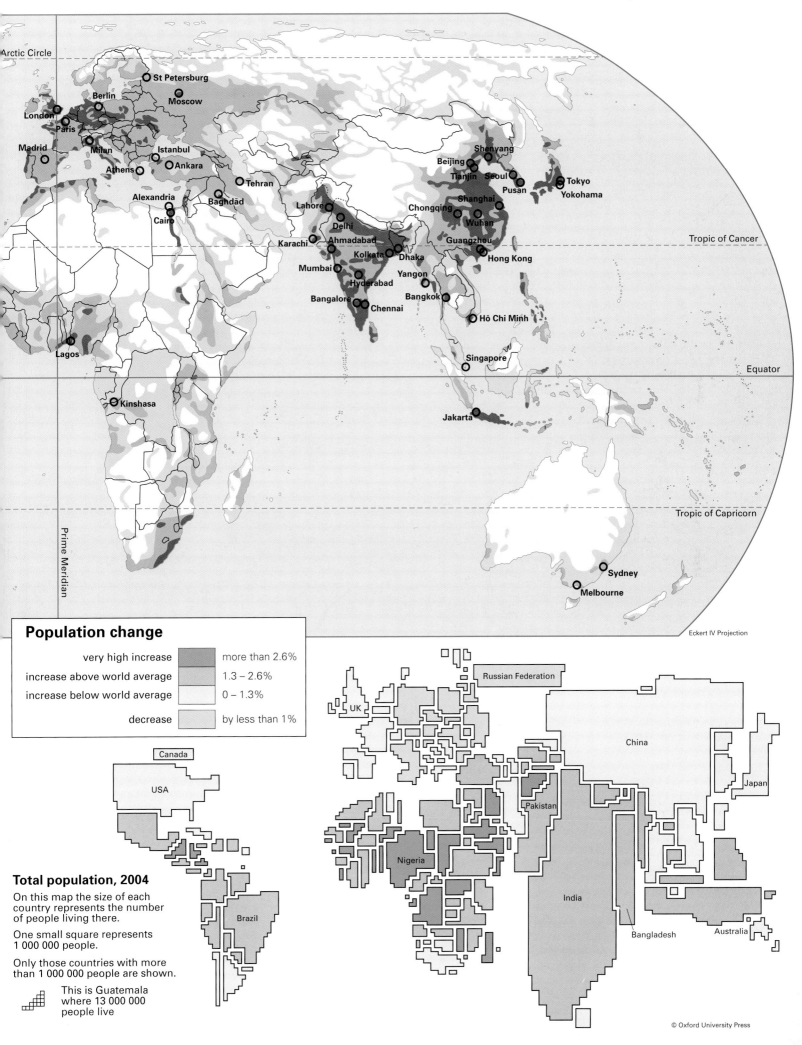

Population change

very high increase		more than 2.6%
increase above world average		1.3 – 2.6%
increase below world average		0 – 1.3%
decrease		by less than 1%

Total population, 2004

On this map the size of each country represents the number of people living there.

One small square represents 1 000 000 people.

Only those countries with more than 1 000 000 people are shown.

This is Guatemala where 13 000 000 people live

Eckert IV Projection

© Oxford University Press

Purchasing power

Purchasing Power Parity (PPP), 2002 in $ US

Based on Gross Domestic Product (GDP) per person, adjusted for the local cost of living.

- over 25 000
- 10 000–25 000
- 5000–10 000
- 2500–5000
- 1000–2500
- under 1000
- data not available

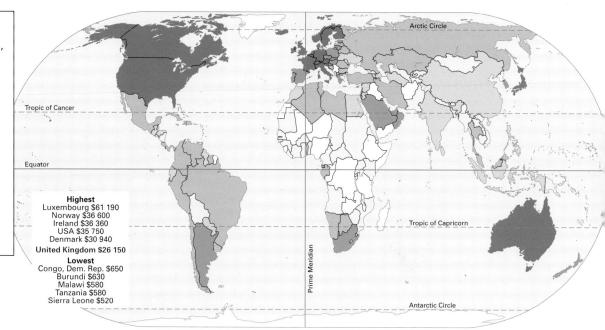

Highest
Luxembourg $61 190
Norway $36 600
Ireland $36 360
USA $35 750
Denmark $30 940
United Kingdom $26 150
Lowest
Congo, Dem. Rep. $650
Burundi $630
Malawi $580
Tanzania $580
Sierra Leone $520

Givers and receivers of aid, 2002 in $ US

Givers
- over $100 per person
- $50–$100 per person
- under $50 per person
- **countries neither giving nor receiving**

Receivers
- under $10 per person
- $10–$100 per person
- over $100 per person
- data not available

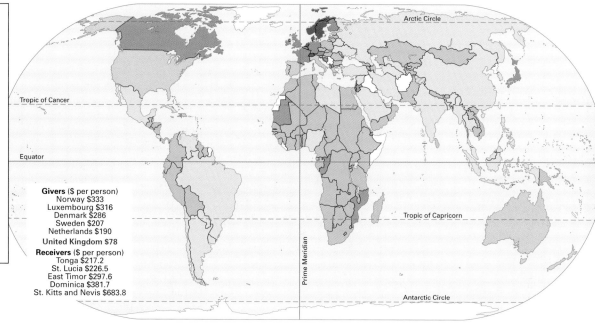

Givers ($ per person)
Norway $333
Luxembourg $316
Denmark $286
Sweden $207
Netherlands $190
United Kingdom $78
Receivers ($ per person)
Tonga $217.2
St. Lucia $226.5
East Timor $297.6
Dominica $381.7
St. Kitts and Nevis $683.8

Life expectancy

Average number of years a baby born in 2002 can expect to live

- over 70 years
- 65–70 years
- 60–65 years
- 55–60 years
- under 55 years
- data not available

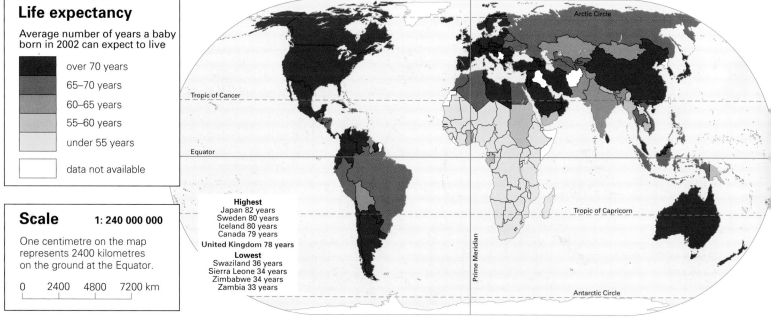

Highest
Japan 82 years
Sweden 80 years
Iceland 80 years
Canada 79 years
United Kingdom 78 years
Lowest
Swaziland 36 years
Sierra Leone 34 years
Zimbabwe 34 years
Zambia 33 years

Scale   1: 240 000 000

One centimetre on the map represents 2400 kilometres on the ground at the Equator.

0   2400   4800   7200 km

Eckert IV Projection

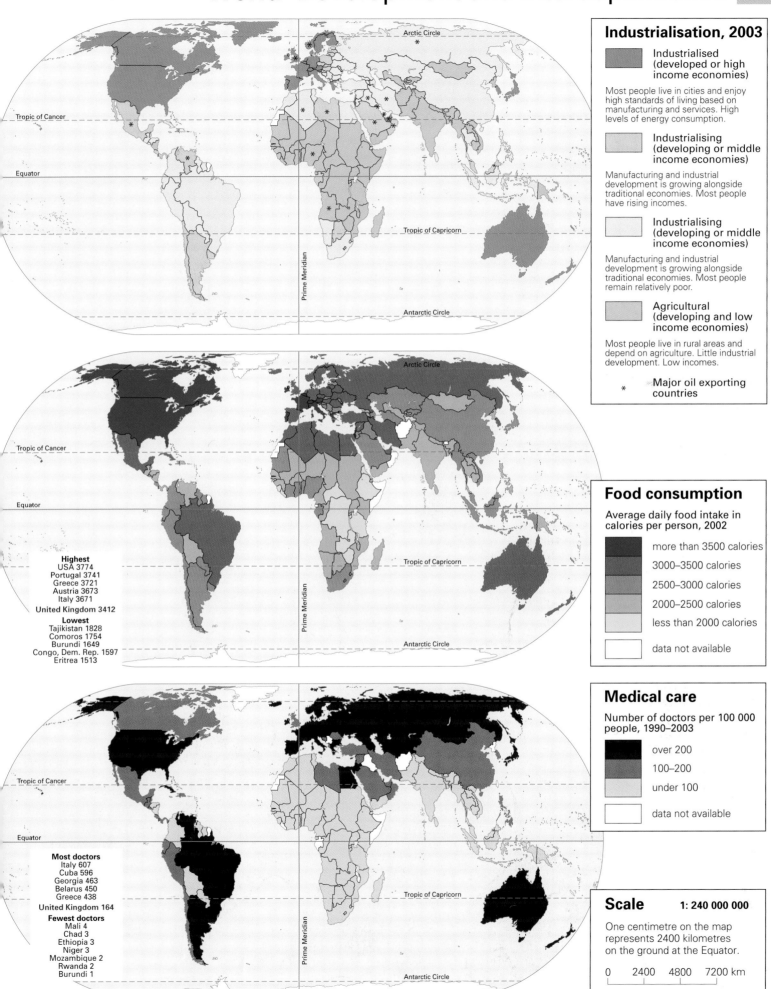

Industrialisation, 2003

Industrialised (developed or high income economies)

Most people live in cities and enjoy high standards of living based on manufacturing and services. High levels of energy consumption.

Industrialising (developing or middle income economies)

Manufacturing and industrial development is growing alongside traditional economies. Most people have rising incomes.

Industrialising (developing or middle income economies)

Manufacturing and industrial development is growing alongside traditional economies. Most people remain relatively poor.

Agricultural (developing and low income economies)

Most people live in rural areas and depend on agriculture. Little industrial development. Low incomes.

* Major oil exporting countries

Highest
USA 3774
Portugal 3741
Greece 3721
Austria 3673
Italy 3671
United Kingdom 3412
Lowest
Tajikistan 1828
Comoros 1754
Burundi 1649
Congo, Dem. Rep. 1597
Eritrea 1513

Food consumption

Average daily food intake in calories per person, 2002

more than 3500 calories

3000–3500 calories

2500–3000 calories

2000–2500 calories

less than 2000 calories

data not available

Medical care

Number of doctors per 100 000 people, 1990–2003

over 200

100–200

under 100

data not available

Most doctors
Italy 607
Cuba 596
Georgia 463
Belarus 450
Greece 438
United Kingdom 164
Fewest doctors
Mali 4
Chad 3
Ethiopia 3
Niger 3
Mozambique 2
Rwanda 2
Burundi 1

Scale   1: 240 000 000

One centimetre on the map represents 2400 kilometres on the ground at the Equator.

0   2400   4800   7200 km

Eckert IV Projection   © Oxford University Press

Water

Surplus

Enough water to support vegetation and crops without irrigation.

- large surplus
- surplus

Deficiency

Not enough water to support vegetation and crops without irrigation. After long periods of deficiency, these areas may lose their natural vegetation.

- deficiency
- chronic deficiency

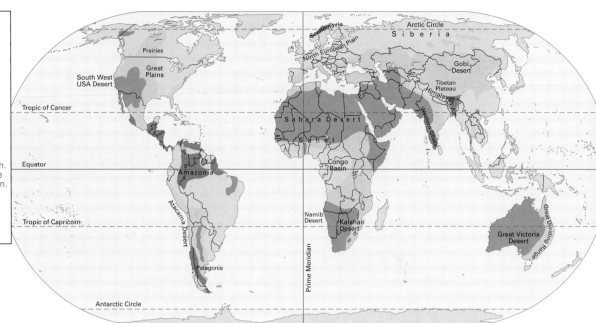

Desertification

- existing areas of desert
- areas with a high risk of desertification
- areas with a moderate risk of desertification

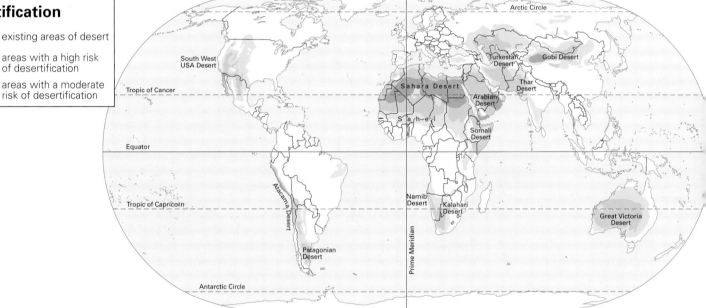

Tropical deforestation

- existing areas of rainforest
- former areas of rainforest

Countries losing greatest areas of forest (000 hectares per year) 1990–2000
Brazil 2309
Indonesia 1312
Sudan 959
Zambia 851
Mexico 631
Congo, Democratic Republic 532
Myanmar 517

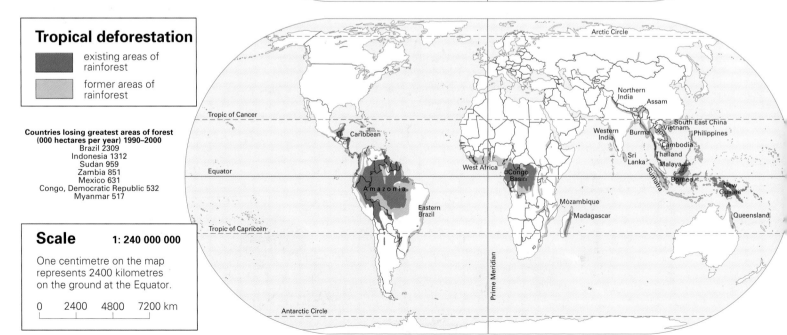

Scale   1: 240 000 000

One centimetre on the map represents 2400 kilometres on the ground at the Equator.

0   2400   4800   7200 km

Eckert IV Projection
© Oxford University Press

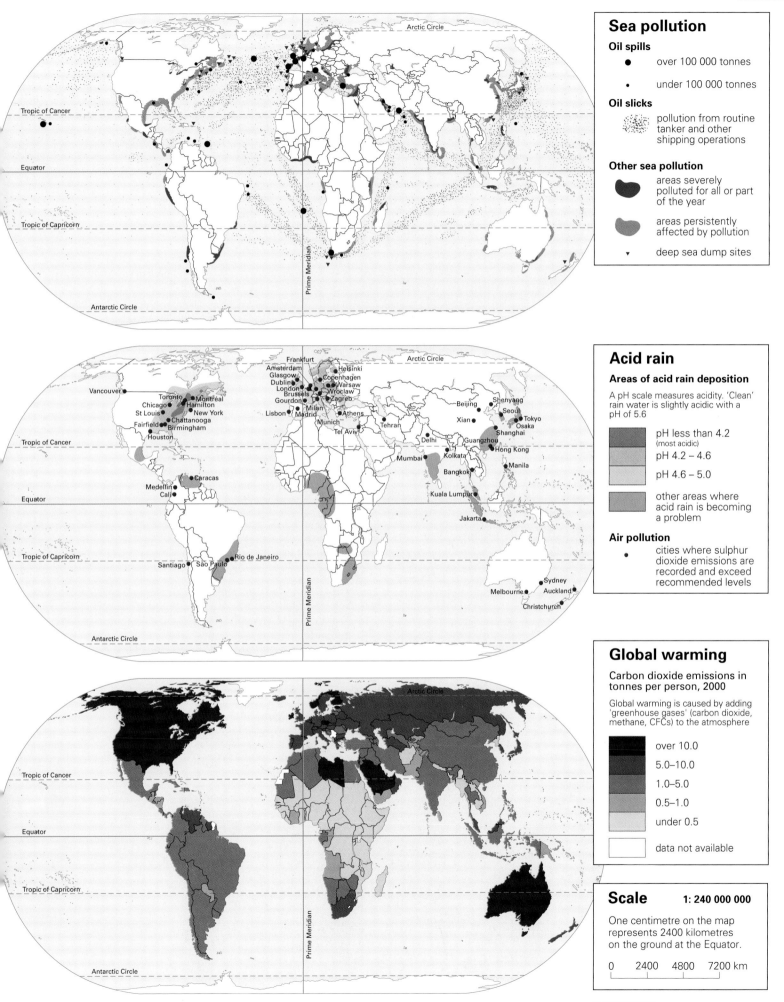

World Environment and Pollution 17

Sea pollution

Oil spills
- ● over 100 000 tonnes
- • under 100 000 tonnes

Oil slicks
pollution from routine tanker and other shipping operations

Other sea pollution
areas severely polluted for all or part of the year

areas persistently affected by pollution

▼ deep sea dump sites

Acid rain

Areas of acid rain deposition
A pH scale measures acidity. 'Clean' rain water is slightly acidic with a pH of 5.6

- pH less than 4.2 (most acidic)
- pH 4.2 – 4.6
- pH 4.6 – 5.0
- other areas where acid rain is becoming a problem

Air pollution
• cities where sulphur dioxide emissions are recorded and exceed recommended levels

Global warming

Carbon dioxide emissions in tonnes per person, 2000
Global warming is caused by adding 'greenhouse gases' (carbon dioxide, methane, CFCs) to the atmosphere

- over 10.0
- 5.0–10.0
- 1.0–5.0
- 0.5–1.0
- under 0.5
- data not available

Scale   1: 240 000 000
One centimetre on the map represents 2400 kilometres on the ground at the Equator.

0   2400   4800   7200 km

Eckert IV Projection

© Oxford University Press

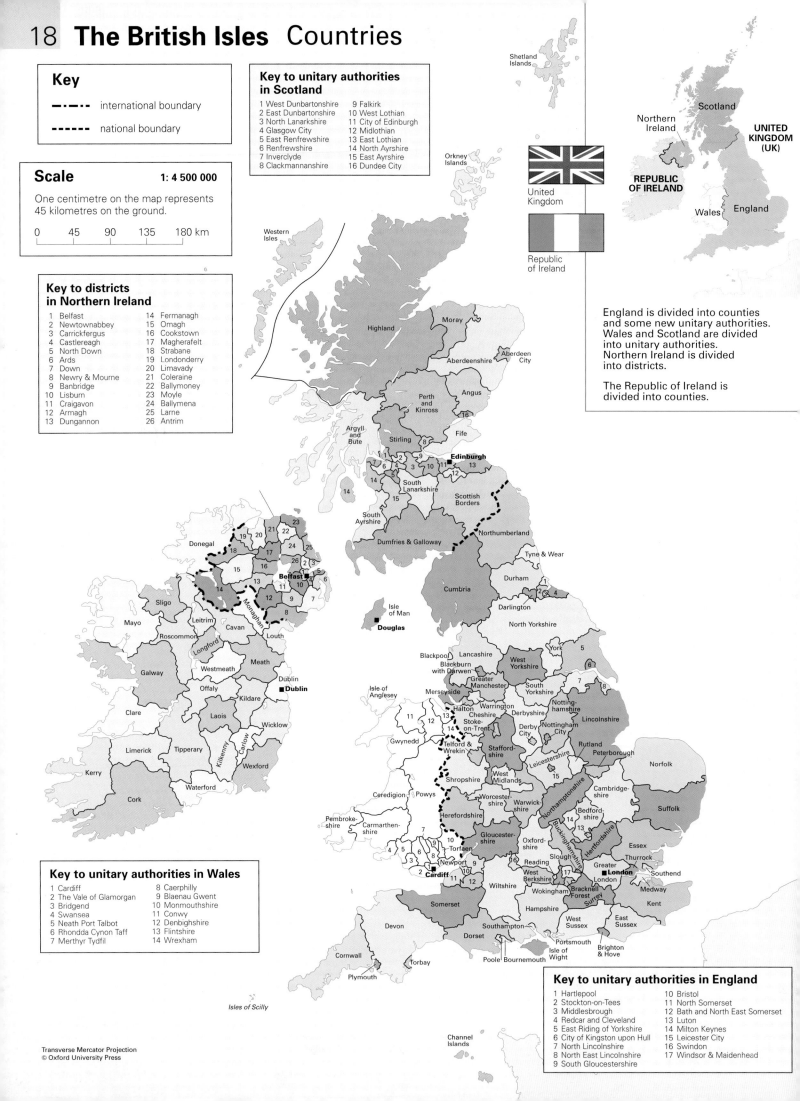

Key

—·—·— international boundary

— — — — national boundary

Scale

1: 4 500 000

One centimetre on the map represents
45 kilometres on the ground.

0   45   90   135   180 km

Key to unitary authorities in Scotland

1 West Dunbartonshire
2 East Dunbartonshire
3 North Lanarkshire
4 Glasgow City
5 East Renfrewshire
6 Renfrewshire
7 Inverclyde
8 Clackmannanshire
9 Falkirk
10 West Lothian
11 City of Edinburgh
12 Midlothian
13 East Lothian
14 North Ayrshire
15 East Ayrshire
16 Dundee City

Key to districts in Northern Ireland

1 Belfast
2 Newtownabbey
3 Carrickfergus
4 Castlereagh
5 North Down
6 Ards
7 Down
8 Newry & Mourne
9 Banbridge
10 Lisburn
11 Craigavon
12 Armagh
13 Dungannon
14 Fermanagh
15 Omagh
16 Cookstown
17 Magherafelt
18 Strabane
19 Londonderry
20 Limavady
21 Coleraine
22 Ballymoney
23 Moyle
24 Ballymena
25 Larne
26 Antrim

United Kingdom

Republic of Ireland

England is divided into counties
and some new unitary authorities.
Wales and Scotland are divided
into unitary authorities.
Northern Ireland is divided
into districts.

The Republic of Ireland is
divided into counties.

UNITED KINGDOM (UK)

Scotland

Northern Ireland

REPUBLIC OF IRELAND

Wales

England

Key to unitary authorities in Wales

1 Cardiff
2 The Vale of Glamorgan
3 Bridgend
4 Swansea
5 Neath Port Talbot
6 Rhondda Cynon Taff
7 Merthyr Tydfil
8 Caerphilly
9 Blaenau Gwent
10 Monmouthshire
11 Conwy
12 Denbighshire
13 Flintshire
14 Wrexham

Key to unitary authorities in England

1 Hartlepool
2 Stockton-on-Tees
3 Middlesbrough
4 Redcar and Cleveland
5 East Riding of Yorkshire
6 City of Kingston upon Hull
7 North Lincolnshire
8 North East Lincolnshire
9 South Gloucestershire
10 Bristol
11 North Somerset
12 Bath and North East Somerset
13 Luton
14 Milton Keynes
15 Leicester City
16 Swindon
17 Windsor & Maidenhead

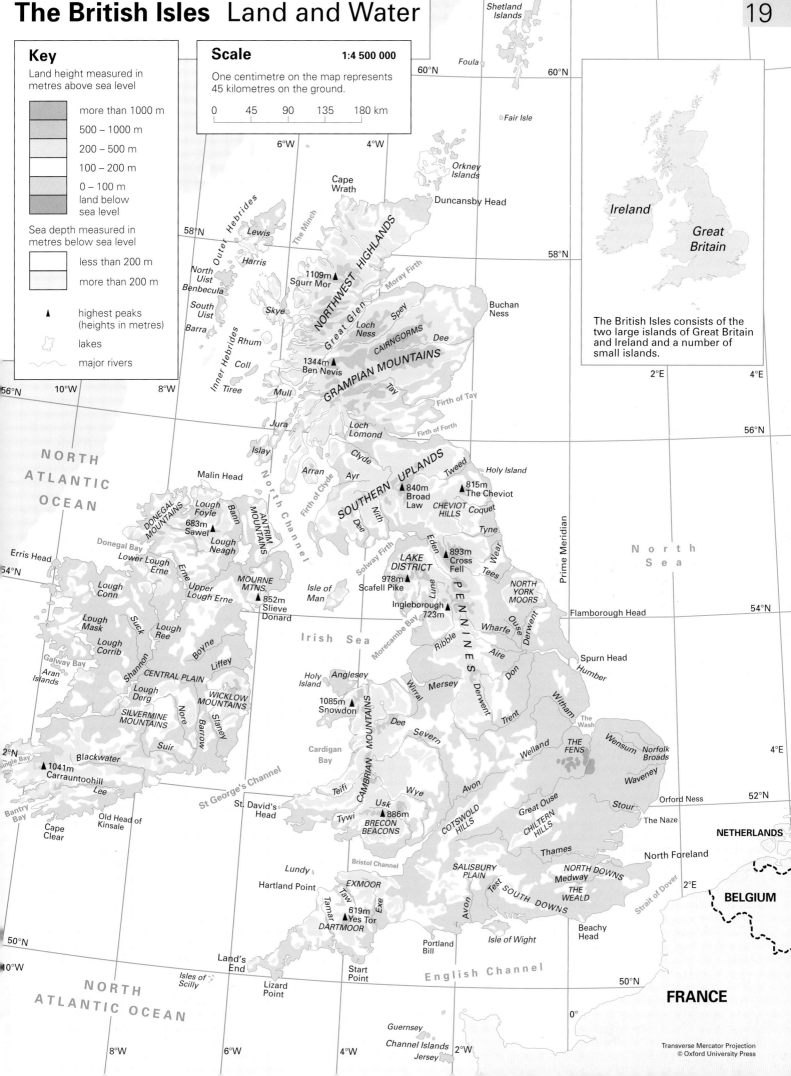

Key

Land height measured in metres above sea level

- more than 1000 m
- 500 – 1000 m
- 200 – 500 m
- 100 – 200 m
- 0 – 100 m
- land below sea level

Sea depth measured in metres below sea level

- less than 200 m
- more than 200 m

- ▲ highest peaks (heights in metres)
- lakes
- major rivers

Scale

1:4 500 000

One centimetre on the map represents 45 kilometres on the ground.

| 0 | 45 | 90 | 135 | 180 km |

The British Isles consists of the two large islands of Great Britain and Ireland and a number of small islands.

Ireland *Great Britain*

Transverse Mercator Projection
© Oxford University Press

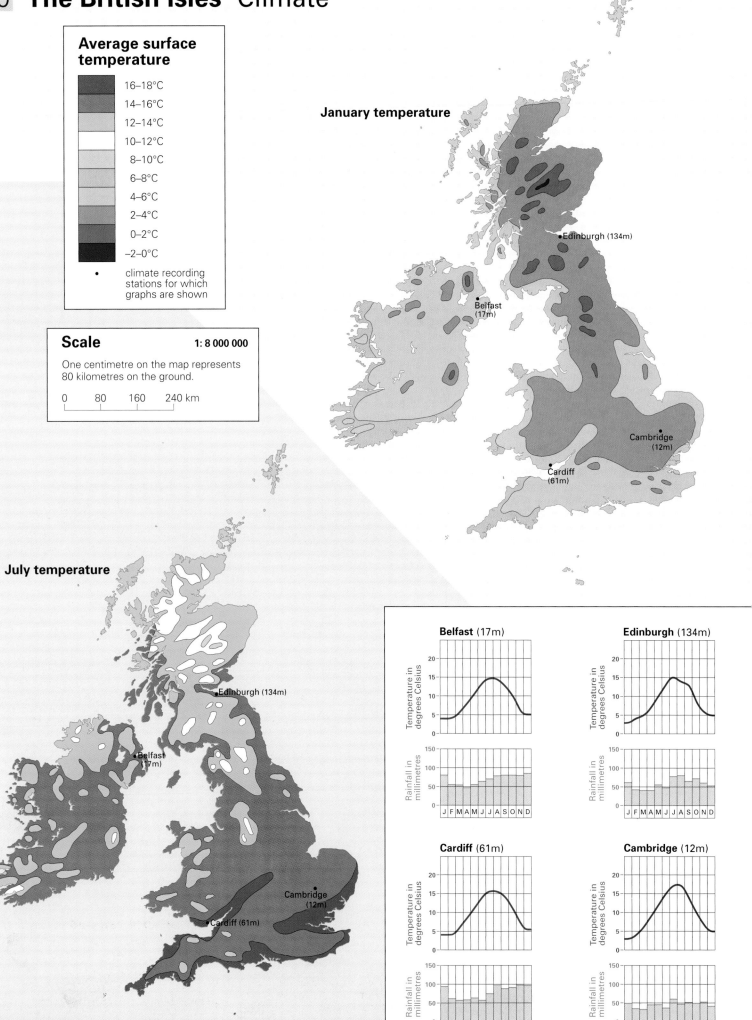

Average surface temperature

	16–18°C
	14–16°C
	12–14°C
	10–12°C
	8–10°C
	6–8°C
	4–6°C
	2–4°C
	0–2°C
	–2–0°C
•	climate recording stations for which graphs are shown

Scale   1: 8 000 000

One centimetre on the map represents 80 kilometres on the ground.

0   80   160   240 km

January temperature

•Edinburgh (134m)

•Belfast (17m)

Cambridge (12m)

Cardiff (61m)

July temperature

•Edinburgh (134m)

•Belfast (17m)

Cambridge (12m)

•Cardiff (61m)

Belfast (17m)

Temperature in degrees Celsius

Rainfall in millimetres

J F M A M J J A S O N D

Edinburgh (134m)

Temperature in degrees Celsius

Rainfall in millimetres

J F M A M J J A S O N D

Cardiff (61m)

Temperature in degrees Celsius

Rainfall in millimetres

J F M A M J J A S O N D

Cambridge (12m)

Temperature in degrees Celsius

Rainfall in millimetres

J F M A M J J A S O N D

Average annual rainfall

more than 2400 millimetres

1200 – 2400 millimetres

800 – 1200 millimetres

less than 800 millimetres

• climate recording stations for which graphs are shown

Drought and flood

inland areas in regular danger of flooding

coastal areas in regular danger of flooding

areas in regular danger of drought

Scale   1: 8 000 000

One centimetre on the map measures 80 kilometres on the ground.

0   80   160   240 km

Edinburgh (134m)

•Belfast (17m)

Cambridge (12m)

Cardiff (61m)

Scale   1: 16 000 000

One centimetre on the map represents 160 kilometres on the ground.

0   160   320   480 km

The water cycle

precipitation

clouds

condensation

vaporation

rain

snow

ice

lake

groundwater

river

sea

Arrows show movement of water or change from one state to another.

Cold winters, cool summers

Mild winters, cool summers

Cool winters, warm summers

Mild winters, warm summers

Climate regions

- - - - -   average January temperature (4°C)

———   average July temperature (16°C)

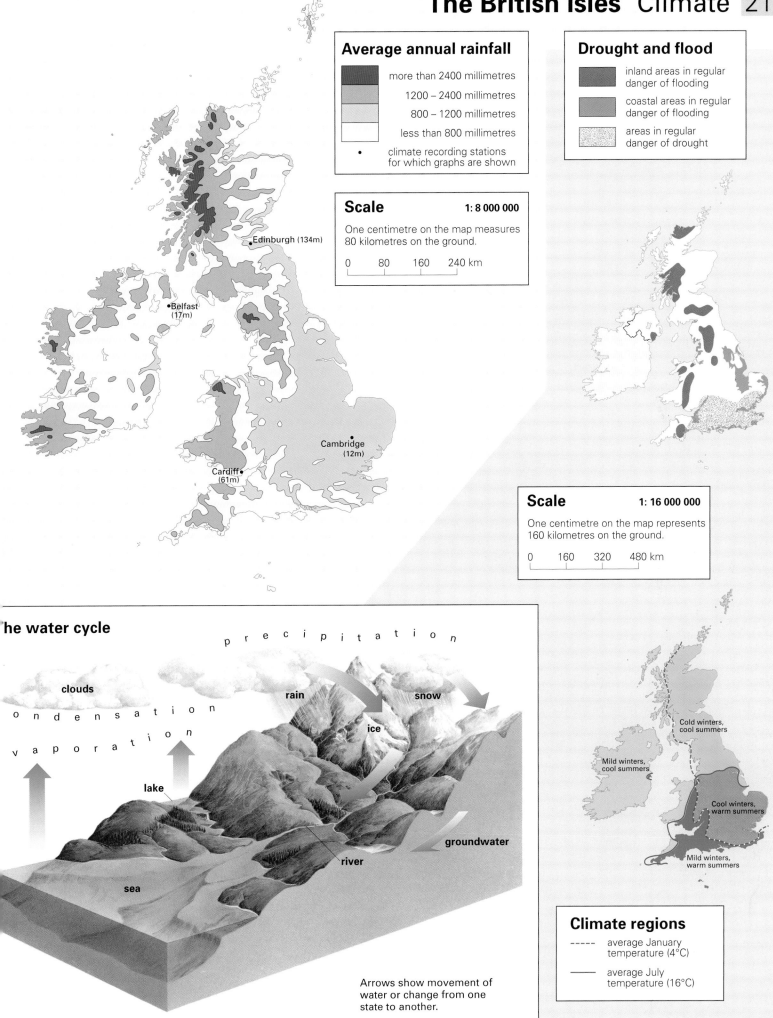

Population structure of the United Kingdom

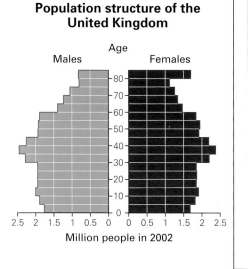

Age

Males   Females

2.5 1.5 1 0.5 0   0 0.5 1 1.5 2 2.5
      Million people in 2002

Population density, 2002

- ■ more than 1000 people per square kilometre
- ■ 500–1000 people per square kilometre
- ■ 100–500 people per square kilometre
- □ less than 100 person per square kilometre

- – – – international boundary
- ——— national boundary
- ——— county, unitary authority, or district boundary

Major cities

- ● with more than 6 million people
- ● with 1 million people
- • with between 400 000 and 1 million people
- · with between 100 000 and 400 000 people

Scale   1: 8 000 000

One centimetre on the map represents 80 kilometres on the ground.

0   80   160   240 km

British Isles population data

| United Kingdom | Overall population density 244 people per square kilometre |
| Republic of Ireland | Overall population density 55 people per square kilometre |

Total population 2002

England	50.0 million people
Wales	2.9 million people
Scotland	5.1 million people
Northern Ireland	1.7 million people
United Kingdom	59.7 million people
Republic of Ireland	3.9 million people

Population change

Change in population in each county, region or district, 1982 – 2002

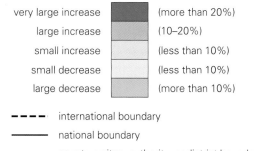

very large increase	(more than 20%)
large increase	(10–20%)
small increase	(less than 10%)
small decrease	(less than 10%)
large decrease	(more than 10%)

- – – – international boundary
- ——— national boundary
- ——— county, unitary authority, or district boundary

Transverse Mercator Projection
© Oxford University Press

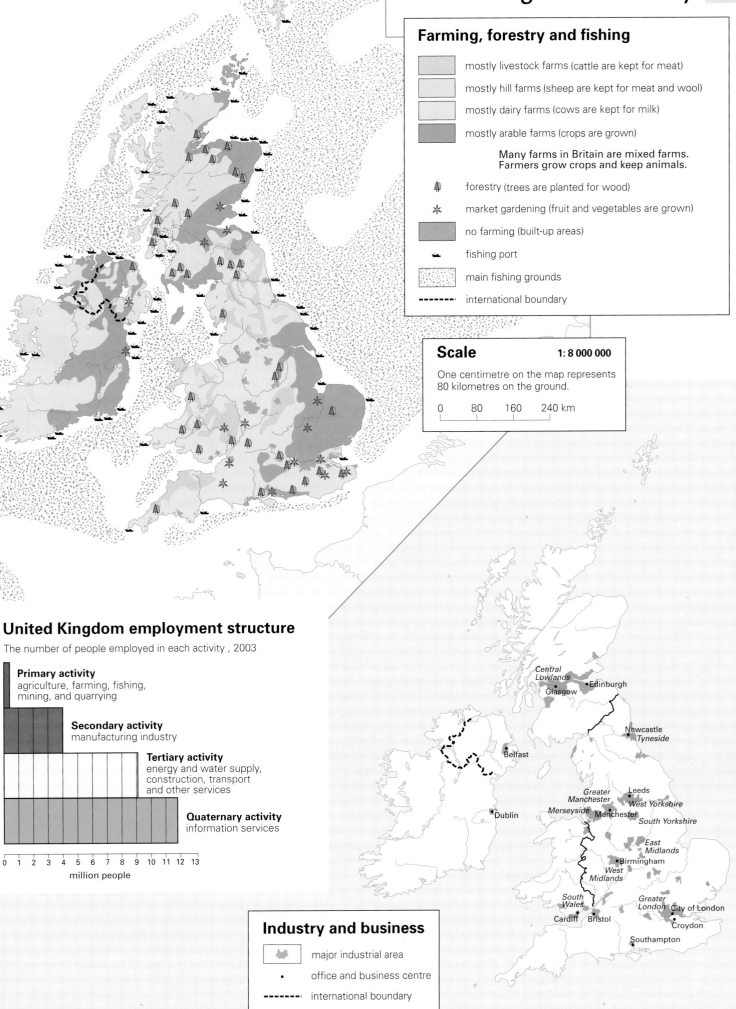

Farming, forestry and fishing

mostly livestock farms (cattle are kept for meat)

mostly hill farms (sheep are kept for meat and wool)

mostly dairy farms (cows are kept for milk)

mostly arable farms (crops are grown)

Many farms in Britain are mixed farms. Farmers grow crops and keep animals.

forestry (trees are planted for wood)

market gardening (fruit and vegetables are grown)

no farming (built-up areas)

fishing port

main fishing grounds

- - - - international boundary

Scale   1: 8 000 000

One centimetre on the map represents 80 kilometres on the ground.

0   80   160   240 km

United Kingdom employment structure

The number of people employed in each activity , 2003

Primary activity
agriculture, farming, fishing, mining, and quarrying

Secondary activity
manufacturing industry

Tertiary activity
energy and water supply, construction, transport and other services

Quaternary activity
information services

0 1 2 3 4 5 6 7 8 9 10 11 12 13
million people

Industry and business

major industrial area

• office and business centre

- - - - international boundary

——— national boundary

Central Lowlands
Glasgow
Edinburgh
Newcastle
Tyneside
Belfast
Greater Manchester
Leeds
West Yorkshire
Merseyside
Manchester
South Yorkshire
Dublin
East Midlands
Birmingham
West Midlands
South Wales
Greater London
City of London
Cardiff
Bristol
Croydon
Southampton

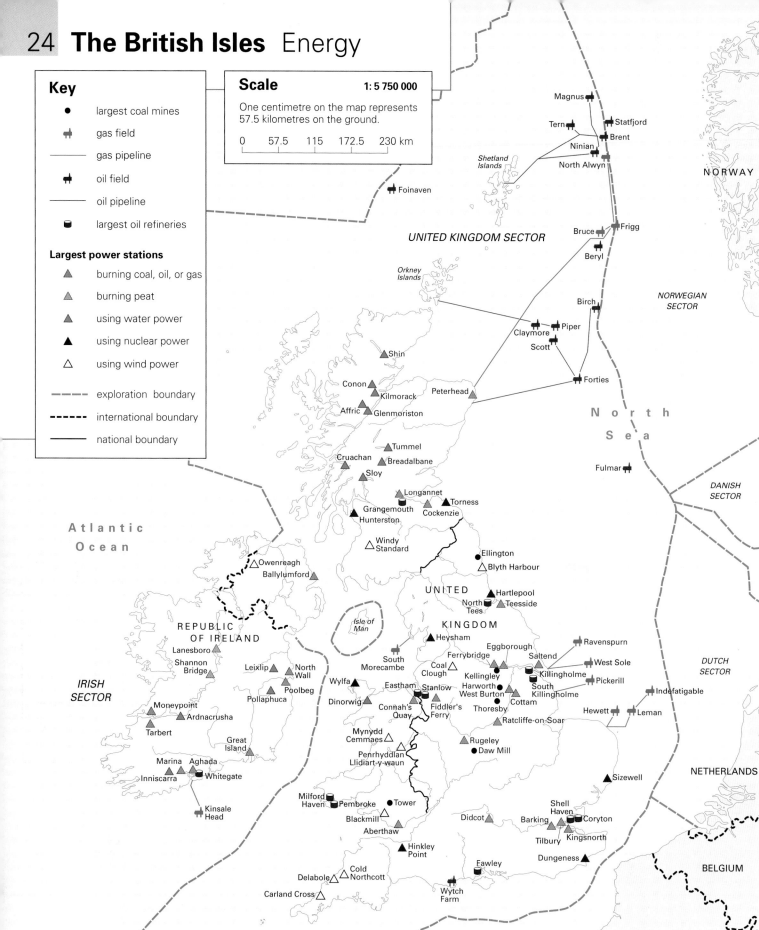

Key

- ● largest coal mines
- ⊶ gas field
- — gas pipeline
- ⊶ oil field
- — oil pipeline
- ▭ largest oil refineries

Largest power stations

- ▲ burning coal, oil, or gas
- ▲ burning peat
- ▲ using water power
- ▲ using nuclear power
- △ using wind power

- – – – exploration boundary
- ····· international boundary
- —— national boundary

Scale

1: 5 750 000

One centimetre on the map represents 57.5 kilometres on the ground.

0   57.5   115   172.5   230 km

NORWAY

Magnus
Tern
Statfjord
Brent
Ninian
North Alwyn
Foinaven

Shetland Islands

UNITED KINGDOM SECTOR

Bruce
Frigg
Beryl

Orkney Islands

NORWEGIAN SECTOR

Birch
Claymore
Piper
Scott

Shin

Conon
Kilmorack
Affric
Glenmoriston

Peterhead
Forties

N o r t h
S e a

Tummel
Cruachan
Breadalbane
Sloy

Longannet
Grangemouth
Hunterston
Cockenzie
Torness

Fulmar

DANISH SECTOR

A t l a n t i c
O c e a n

Windy Standard

Owenreagh
Ballylumford

Ellington
Blyth Harbour

UNITED

Hartlepool
North Tees
Teesside

REPUBLIC
OF IRELAND

Isle of Man

KINGDOM

Heysham

Lanesboro
Shannon Bridge

Leixlip
North Wall
Poolbeg
Pollaphuca

South Morecambe

Ferrybridge
Eggborough
Coal Clough
Saltend
Killingholme
West Sole

Ravenspurn

DUTCH SECTOR

IRISH
SECTOR

Wylfa

Eastham
Stanlow
Dinorwig
Connah's Quay
Fiddler's Ferry

Kellingley
Harworth
West Burton
Thoresby
Cottam
South Killingholme

Pickerill
Indefatigable

Moneypoint
Ardnacrusha
Tarbert

Great Island

Mynydd Cemmaes

Ratcliffe-on-Soar

Hewett
Leman

Marina
Aghada
Inniscarra
Whitegate

Penrhyddlan
Llidiart-y-waun

Rugeley
Daw Mill

Sizewell

NETHERLANDS

Kinsale Head

Milford Haven
Pembroke
Blackmill
Aberthaw

Tower

Didcot

Shell Haven
Barking
Tilbury
Coryton
Kingsnorth

Hinkley Point

Delabole
Cold Northcott

Fawley

Dungeness

Carland Cross

Wytch Farm

BELGIUM

A t l a n t i c
O c e a n

Channel Islands

F R A N C E

Transverse Mercator Projection
© Oxford University Press

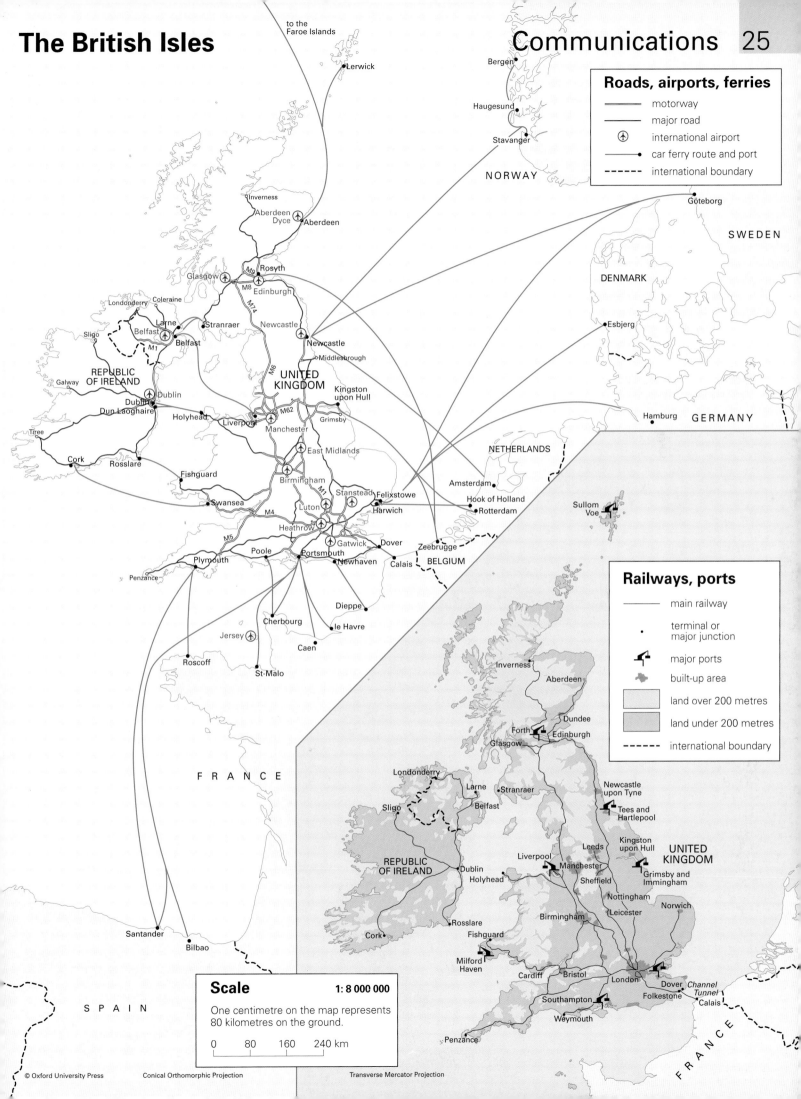

The British Isles

to the
Faroe Islands

Roads, airports, ferries

══════	motorway
──────	major road
⊛	international airport
●─────	car ferry route and port
─ ─ ─ ─	international boundary

Lerwick

Bergen

Haugesund

Stavanger

NORWAY

SWEDEN

Göteborg

DENMARK

Inverness

Aberdeen
Dyce ⊛ Aberdeen

Esbjerg

Rosyth

Glasgow ⊛ M9
M8 Edinburgh

M74

Londonderry
Coleraine

Larne
Belfast ⊛ Stranraer
M1 Belfast

Sligo

Newcastle
⊛ Newcastle

Middlesbrough

Hamburg GERMANY

REPUBLIC
OF IRELAND

Galway

UNITED
KINGDOM

Kingston
upon Hull

M6

NETHERLANDS

Dublin ⊛
Dun Laoghaire

Dublin

Holyhead

Liverpool
Manchester M62

Grimsby

Amsterdam

Hook of Holland
Rotterdam

Sullom
Voe

Tiree

Cork

Rosslare

Fishguard

East Midlands ⊛

Birmingham ⊛

Stanstead ⊛
Felixstowe

Swansea

M1

Luton ⊛

Harwich

M4

Heathrow ⊛

Gatwick ⊛ Dover

Zeebrugge

BELGIUM

M5

Poole

Portsmouth

Newhaven

Calais

Plymouth

Penzance

Dieppe

le Havre

Jersey ⊛

Cherbourg

Caen

Roscoff

St-Malo

FRANCE

Railways, ports

──────	main railway
●	terminal or major junction
⚑	major ports
▨	built-up area
░	land over 200 metres
▒	land under 200 metres
─ ─ ─ ─	international boundary

Inverness

Aberdeen

Dundee

Forth ⚑
Glasgow Edinburgh

Londonderry

Larne
Stranraer

Newcastle
upon Tyne

Tees and
Hartlepool

Sligo

Belfast

REPUBLIC
OF IRELAND

Liverpool
Manchester

Leeds

Sheffield

Kingston
upon Hull

Grimsby and
Immingham

UNITED
KINGDOM

Dublin
Holyhead

Nottingham

Leicester

Norwich

Rosslare

Birmingham

Cork

Fishguard

Milford
Haven

Cardiff

Bristol

London

Dover Channel
Tunnel

Folkestone Calais

Southampton

Weymouth

Penzance

FRANCE

Santander

Bilbao

SPAIN

Scale

1: 8 000 000

One centimetre on the map represents
80 kilometres on the ground.

0   80   160   240 km

© Oxford University Press   Conical Orthomorphic Projection   Transverse Mercator Projection

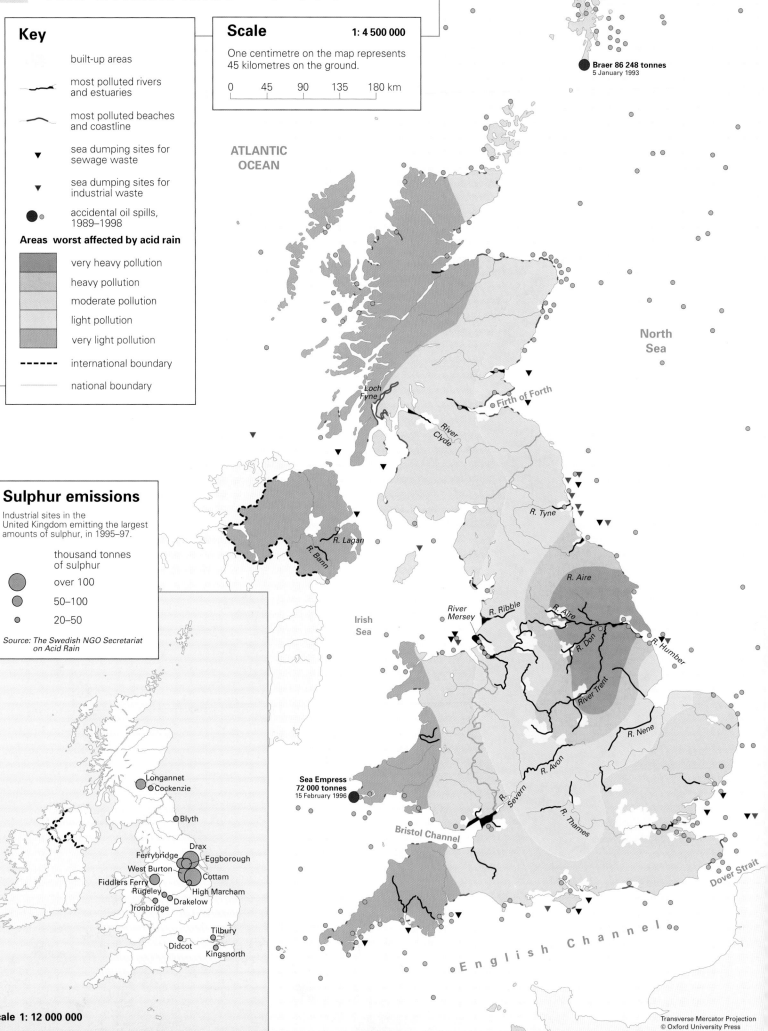

Key

- built-up areas
- most polluted rivers and estuaries
- most polluted beaches and coastline
- ▼ sea dumping sites for sewage waste
- ▼ sea dumping sites for industrial waste
- ● ● accidental oil spills, 1989–1998

Areas worst affected by acid rain

- very heavy pollution
- heavy pollution
- moderate pollution
- light pollution
- very light pollution
- - - - international boundary
- national boundary

Scale

1: 4 500 000

One centimetre on the map represents 45 kilometres on the ground.

| 0 | 45 | 90 | 135 | 180 km |

ATLANTIC OCEAN

Braer 86 248 tonnes
5 January 1993

North Sea

Loch Fyne

Firth of Forth

River Clyde

R. Tyne

R. Lagan

R. Bann

R. Aire

River Mersey

R. Ribble

R. Aire

R. Don

R. Humber

Irish Sea

River Trent

R. Nene

Sea Empress 72 000 tonnes
15 February 1996

R. Severn

R. Avon

R. Thames

Bristol Channel

Dover Strait

English Channel

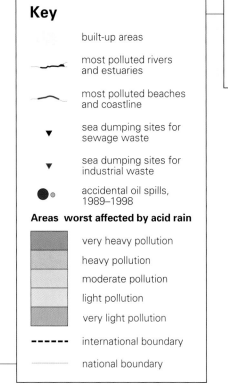

Sulphur emissions

Industrial sites in the United Kingdom emitting the largest amounts of sulphur, in 1995–97.

thousand tonnes of sulphur

- ● over 100
- ● 50–100
- ● 20–50

Source: The Swedish NGO Secretariat on Acid Rain

Longannet
Cockenzie
Blyth
Drax
Ferrybridge
Eggborough
West Burton
Cottam
Fiddlers Ferry
Rugeley
High Marcham
Ironbridge
Drakelow
Tilbury
Didcot
Kingsnorth

Scale 1: 12 000 000

Transverse Mercator Projection
© Oxford University Press

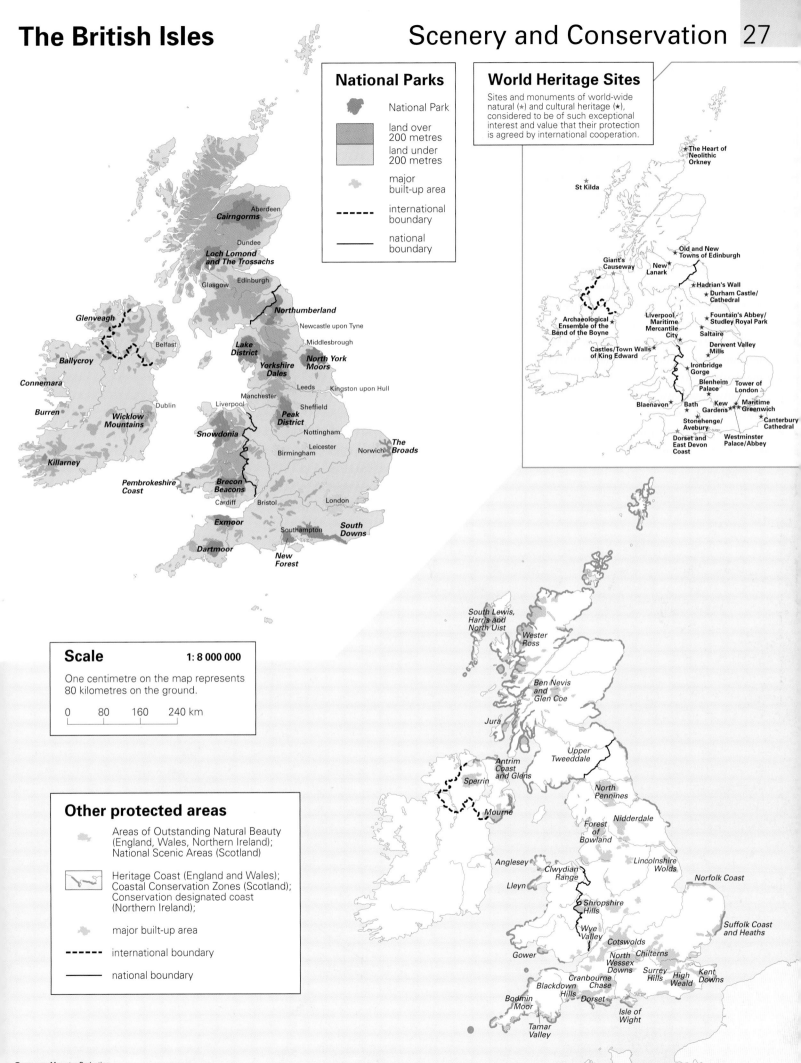

National Parks

- National Park
- land over 200 metres
- land under 200 metres
- major built-up area
- - - - international boundary
- ——— national boundary

World Heritage Sites

Sites and monuments of world-wide natural (✶) and cultural heritage (★), considered to be of such exceptional interest and value that their protection is agreed by international cooperation.

Cairngorms
Aberdeen
Dundee
Loch Lomond and The Trossachs
Glasgow
Edinburgh
Glenveagh
Belfast
Northumberland
Newcastle upon Tyne
Middlesbrough
Ballycroy
Lake District
North York Moors
Connemara
Yorkshire Dales
Leeds
Kingston upon Hull
Burren
Manchester
Liverpool
Sheffield
Dublin
Wicklow Mountains
Snowdonia
Peak District
Nottingham
Killarney
Birmingham
Leicester
Norwich
The Broads
Pembrokeshire Coast
Brecon Beacons
Cardiff
Bristol
London
Exmoor
Southampton
South Downs
Dartmoor
New Forest

✶ The Heart of Neolithic Orkney
✶ St Kilda
Giant's Causeway ✶
New Lanark ★
Old and New Towns of Edinburgh ★
★ Hadrian's Wall
★ Durham Castle/ Cathedral
Archaeological Ensemble of the Bend of the Boyne
Liverpool Maritime Mercantile City
★ Fountain's Abbey/ Studley Royal Park
★ Saltaire
Castles/Town Walls ★ of King Edward
★ Derwent Valley Mills
★ Ironbridge Gorge
Blenheim Palace
Tower of London
Blaenavon ★
Bath ★
Kew Gardens
Maritime Greenwich
Stonehenge/ Avebury
Canterbury Cathedral
Dorset and East Devon Coast
Westminster Palace/Abbey

Scale

1: 8 000 000

One centimetre on the map represents 80 kilometres on the ground.

0   80   160   240 km

Other protected areas

- Areas of Outstanding Natural Beauty (England, Wales, Northern Ireland); National Scenic Areas (Scotland)
- Heritage Coast (England and Wales); Coastal Conservation Zones (Scotland); Conservation designated coast (Northern Ireland);
- major built-up area
- - - - international boundary
- ——— national boundary

South Lewis, Harris and North Uist
Wester Ross
Ben Nevis and Glen Coe
Jura
Antrim Coast and Glens
Upper Tweeddale
Sperrin
North Pennines
Mourne
Nidderdale
Forest of Bowland
Anglesey
Lincolnshire Wolds
Clwydian Range
Norfolk Coast
Lleyn
Shropshire Hills
Suffolk Coast and Heaths
Wye Valley
Cotswolds
Gower
North Wessex Downs
Chilterns
Surrey Hills
Kent Downs
High Weald
Cranbourne Chase
Blackdown Hills
Dorset
Bodmin Moor
Isle of Wight
Tamar Valley

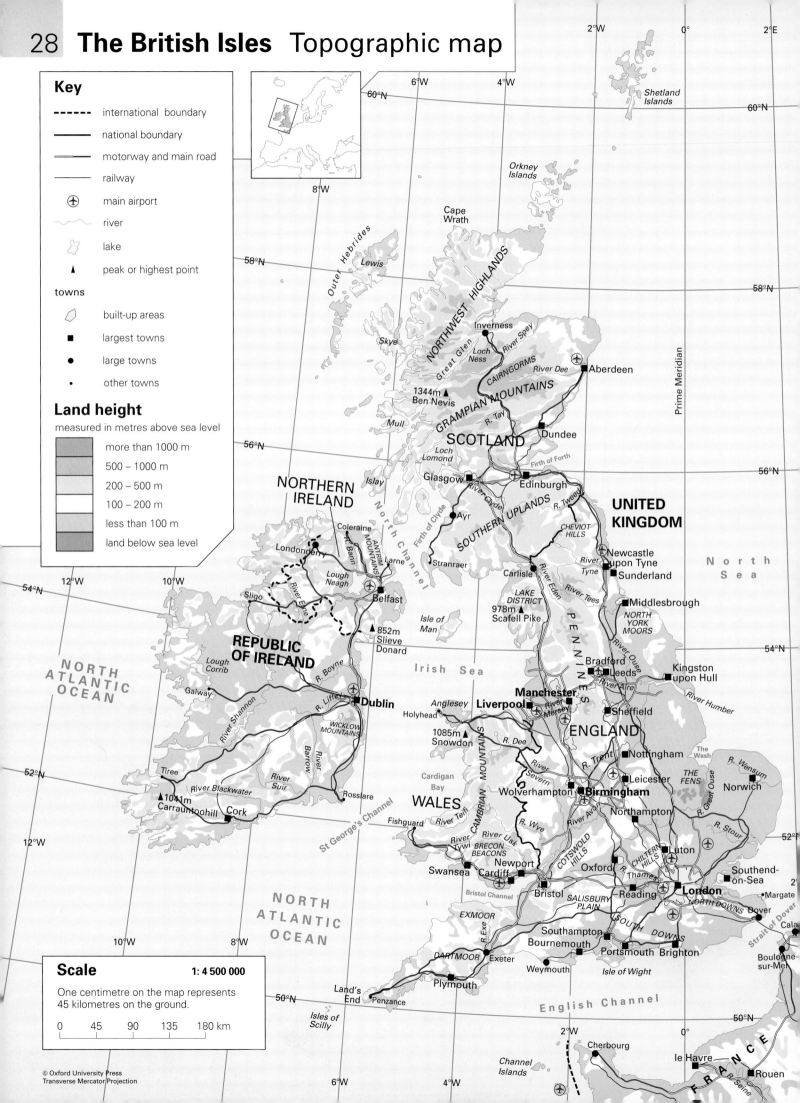

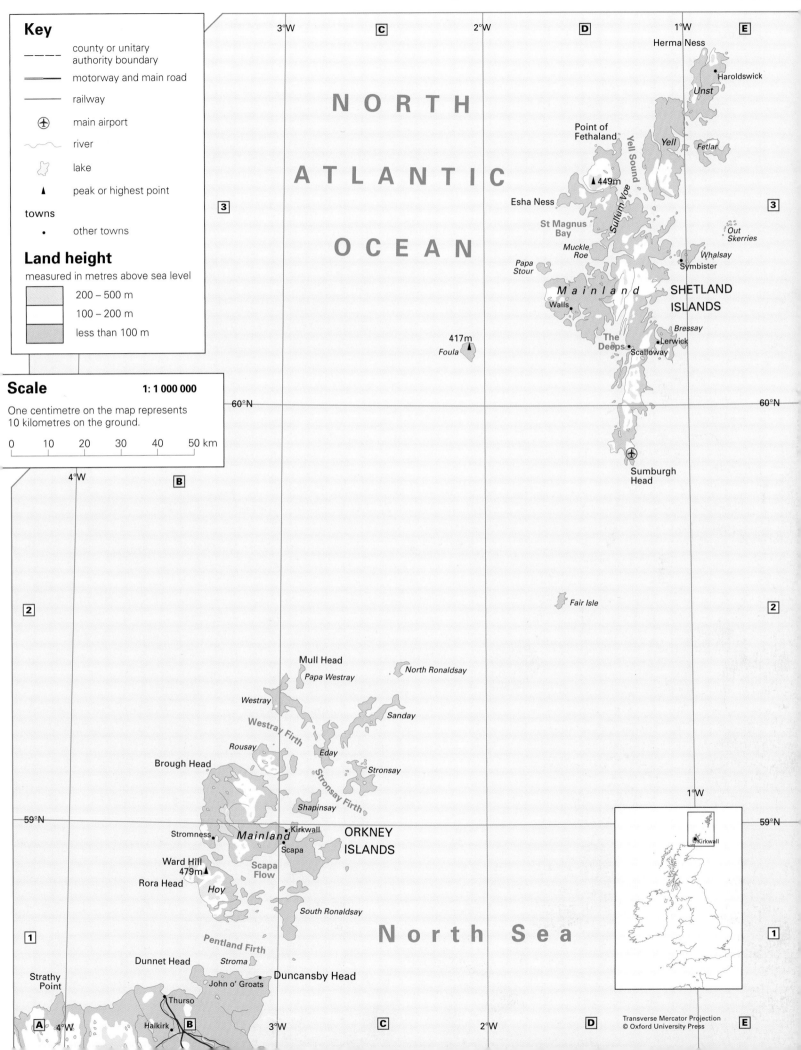

Key

- – – – county or unitary authority boundary
- ═══ motorway and main road
- ─── railway
- ✈ main airport
- ～ river
- 🝝 lake
- ▲ peak or highest point

towns
- • other towns

Land height

measured in metres above sea level

- 200 – 500 m
- 100 – 200 m
- less than 100 m

Scale

1 : 1 000 000

One centimetre on the map represents 10 kilometres on the ground.

0 10 20 30 40 50 km

NORTH ATLANTIC OCEAN

Herma Ness
Haroldswick
Unst
Point of Fethaland
Yell Sound
Yell
Fetlar
Esha Ness
Sullom Voe
St Magnus Bay
▲449m
Out Skerries
Muckle Roe
Whalsay
Papa Stour
Symbister
Mainland
SHETLAND ISLANDS
Walls
Bressay
417m
The Deeps
Lerwick
Foula
Scalloway
60°N   60°N
Sumburgh Head

Fair Isle

Mull Head
Papa Westray
North Ronaldsay
Westray
Sanday
Westray Firth
Rousay
Eday
Brough Head
Stronsay
Stronsay Firth
Shapinsay
Scapa Flow
59°N   59°N
Stromness
Mainland
Kirkwall
ORKNEY ISLANDS
Scapa
Ward Hill 479m▲
Hoy
Rora Head
South Ronaldsay
1°W
North Sea
Pentland Firth
Stroma
Dunnet Head
Strathy Point
Duncansby Head
John o' Groats
Thurso
Halkirk

Kirkwall

Transverse Mercator Projection
© Oxford University Press

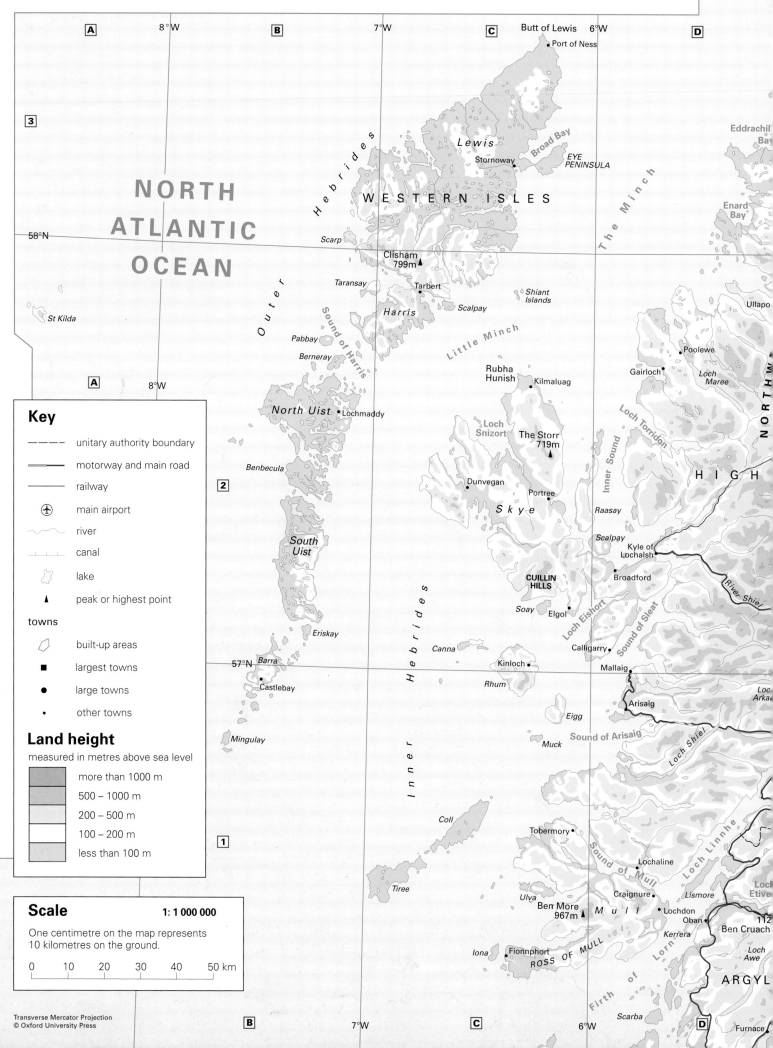

Key

---- unitary authority boundary

motorway and main road

railway

⊕ main airport

river

canal

lake

▲ peak or highest point

towns

built-up areas

■ largest towns

● large towns

· other towns

Land height

measured in metres above sea level

more than 1000 m

500 – 1000 m

200 – 500 m

100 – 200 m

less than 100 m

Scale   1: 1 000 000

One centimetre on the map represents
10 kilometres on the ground.

0   10   20   30   40   50 km

Transverse Mercator Projection
© Oxford University Press

NORTH ATLANTIC OCEAN

St Kilda

Butt of Lewis
Port of Ness

Lewis

Stornoway

Broad Bay

EYE PENINSULA

WESTERN ISLES

Hebrides

The Minch

Eddrachil Bay

Enard Bay

Scarp

Clisham 799m

Taransay

Tarbert

Scalpay

Shiant Islands

Harris

Sound of Harris

Pabbay

Berneray

Outer

Little Minch

Ullapo

Rubha Hunish

Kilmaluag

Poolewe

Gairloch

Loch Maree

North Uist

Lochmaddy

Loch Snizort

The Storr 719m

Loch Torridon

Benbecula

Dunvegan

Portree

Skye

Raasay

Inner Sound

HIGH

South Uist

Scalpay

Kyle of Lochalsh

CUILLIN HILLS

Broadford

River Shiel

Eriskay

Soay

Elgol

Loch Eishort

Canna

Calligarry

Sound of Sleat

Barra

Kinloch

Mallaig

Loch Arka

Castlebay

Rhum

Arisaig

Mingulay

Eigg

Loch Shiel

Muck

Sound of Arisaig

Inner Hebrides

Coll

Tobermory

Craignure

Ben More 967m

Ulva

Mull

Lismore

Lochaline

Lochdon

Oban

Loch Etive

Ben Cruach

Tiree

Iona

Fionnphort

ROSS OF MULL

Kerrera

Firth of Lorn

Loch Awe

ARGYL

Scarba

Furnace

112

E   4°W   F   Dunnet   *Stroma*   3°W
Cape Wrath   Strathy   Head   John o'   Groats
Point

Thurso

Halkirk

Loch Eriboll

Ben Hope
927m   Wick

River Thurso

Loch nan Clar   Kinbrace   Morven ▲705m   Lybster

961m ▲
Ben Klibreck

**North
Sea**

847m
Canisp   ▲ 998m
Ben More
Assynt

Loch Shin

River Helmsdale

Helmsdale

Lairg

Beinn Dearg
1081m   Brora

Bonar Bridge

Dornoch
Dornoch Firth   Tarbat Ness

▲1109m
Sgurr Mór   ▲1046m
Ben Wyvis   Tain

Invergordon   *Cromarty Firth*

River Meig   Cromarty   *Moray Firth*

Dingwall

Branderburgh   Portknockie   Rosehearty   Fraserburgh
Lossiemouth   Portsoy   Banff
R. Spey   Buckie   Cullen   Macduff
Burghead
Elgin   Fochabers
Forres   Aberchirder   Turriff   Peterhead
Nairn   Keith   *River Deveron*   Buchan Ness

R. Beauly   Rothes
Inverness   Charlestown   Huntly
of Aberlour
River Nairn   Dufftown   Ellon
ND

rn Eige
83m   Drumnadrochit   Grantown-
on-Spey   Oldmeldrum
Loch
llardoch   *River Spey*   **M O R A Y**   Inverurie
Loch Ness   *River Don*
Invermoriston   Aviemore   ✈Dyce   **ABERDEEN
CITY**
A B E R D E E N S H I R E   **Aberdeen**
Fort   **MONADHLIATH**
Augustus   **MOUNTAINS**   **CAIRNGORMS**
Kingussie   ▲1244m   *River Dee*
Invergarry   Newtonmore   Cairn Gorm   Aboyne   Banchory
Braemar   Ballater

Loch Lochy   Stonehaven
▲1155m
Ben Alder   Lochnagar   *River North Esk*
William   1148m   **A N G U S**   Laurencekirk
▲1344m   *Loch Ericht*
Ben Nevis   **P E R T H**   Inverbervie
River Isla
Blackwater Reservoir   **A N D**   Milton Ness
Pitlochry   Brechin   Montrose
Loch Rannoch   **K I N R O S S**   Kirriemuir   *River South Esk*
Ben Lawers   Aberfeldy   Alyth   Forfar   Arbroath
1214m   Blairgowrie
Loch Tay   *River Tay*   Rattray
S I D L A W   Carnoustie
Ben More   Coupar   **H I L L S**
Tyndrum   1174m   Angus   **DUNDEE
CITY**
Dalmally   Crianlarich   *Loch Earn*   **Dundee**
Crieff   *Firth of Tay*
S C O T L A N D   *River Earn*   Perth   St Andrews
BUTE   Newburgh
Loch Katrine   Auchterarder   Cupar   Crail
Ben Lomond   Callander   Auchtermuchty   **F I F E**
araray   974m   Dunblane   **O C H I L   H I L L S**   Glenrothes   Anstruther
Tarbet   **S T I R L I N G**   Kinross   *Loch Leven*   Buckhaven
Loch   *River Forth*   **CLACKMANNAN-
SHIRE**
ND *Lomond*

58°N

57°N

H I G H L A N D S

Key

- –·–·– international boundary
- – – – national boundary
- –··– county, district or unitary authority boundary
- motorway and main road
- railway
- ✈ main airport
- river
- canal
- lake
- ▲ peak or highest point

towns

- built-up areas
- ■ largest towns
- ● large towns
- · other towns

Land height

measured in metres above sea level

- more than 1000 m
- 500 – 1000 m
- 200 – 500 m
- 100 – 200 m
- less than 100 m

Transverse Mercator Projection
© Oxford University Press

Map labels

Iona
Fionnphort ROSS OF MULL
Kerrera
Oban
Dalmally
Loch Awe
Firth of Lorn
ARGYLL AND BUT
Inveraray
Furnace
Colonsay
Scalasaig
Scarba
Sound of Jura
Lochgilphead
Loch Fyne
Dunoon
Firth of Clyd
Oronsay
Jura
Port Askaig
Craighouse
Tarbert
Bute
Tighnabruaich
Rothesay
Islay
Gigha
Ardminish
Clachan
Claonaig
Sound of Bute
Lochranza
Kilmory
Kilbrannan Sound
Goat Fell 874 m ▲
Larg
Portnahaven
Ardbeg
Arran Brodick
Port Ellen
Kintyre
NORTH AYRSHIRE
Mull of Oa

Campbeltown
Ailsa Craig
Gir
Southend
Mull of Kintyre

Malin Head
Tory Island
Tory Sound
Rathlin Island
Rathlin Sound
Fair Head
North Channel
Corsewall Point
Ballantrae

Errigal Mountain 752m ▲
Creeslough
INISHOWEN PENINSULA
▲ Slieve Snaght 615m
Buncrana
Lough Swilly
Lough Foyle
Portrush
Portstewart
Coleraine
MOYLE
Ballycastle
Ballymoney
River Bush
COLERAINE
BALLYMONEY
ANTRIM MOUNTAINS
Stranraer
Kilmacrenan
Limavady
Portpatrick
Letterkenny
R. Swilly
LONDONDERRY
Londonderry
LIMAVADY
Dungiven
River Bann
River Main
BALLYMENA
Ballymena
LARNE Larne
DONEGAL
River Finn
Ballybofey
Strabane
STRABANE
Sawel 683m
Maghera
Lough Beg
Island Magee
Donegal
River Derg
Newtownstewart
MAGHERAFELT
Magherafelt
Randalstown
M22
ANTRIM
CARRICKFERGUS
Carrickfergus
Drummore
NORTHERN IRELAND
529m ▲
Antrim
NEWTOWNABBEY
Newtownabbey
Belfast Lough
Bangor
Donaghadee
Ballyshannon
Lower Lough Erne
OMAGH
Omagh
COOKSTOWN
Cookstown
Lough Neagh
Crumlin
Lough
BELFAST
NORTH DOWN
Newtownards
Mull of Galloway
Lough Derg
Coalisland
Dungannon
Belfast
Lisburn
LISBURN
ARDS
Lough Melvin
FERMANAGH
DUNGANNON
R. Blackwater
CRAIGAVON
M1
Lurgan
CASTLEREAGH
Strangford Lough
ARDS PENINSULA
Enniskillen
Portadown
Craigavon
Dromore
Lough Macnean Upper
Armagh
Banbridge
River Bann
DOWN
Downpatrick
Lough Macnean Lower
Upper Lough Erne
ARMAGH
BANBRIDGE
Shannon
LEITRIM
Clones
MONAGHAN
Newtownhamilton
Slieve Donard 852m ▲
Newcastle
St John's Point
Lough Allen
Lough Oughter
Monaghan
Keady
NEWRY AND MOURNE
Newry
Warrenpoint
Kilkeel
Castleblayney
Crossmaglen
Carlingford Lough

REPUBLIC OF IRELAND

Cavan
LOUTH
Dundalk

Scale 1: 1 000 000

One centimetre on the map represents
10 kilometres on the ground.

0   10   20   30   40   50 km

1174m
Ben More
Loch Earn
4°W
Crieff
River Earn
Perth
3°W
Newburgh
St Andrews
G
Auchterarder
Auchtermuchty
Cupar
Crail
Loch Katrine
Callander
OCHIL HILLS
Kinross
FIFE
Anstruther
Ben Lomond
974m
STIRLING
Dunblane CLACKMANNAN-SHIRE
Glenrothes
Loch Leven
Loch Lomond
River Forth
Alloa
Kirkcaldy
North Berwick
Stirling
Dunfermline
Buckhaven
WEST DUNBARTONSHIRE
CAMPSIE FELLS
FALKIRK
M90
Grangemouth
Bo'ness
Inverkeithing
Firth of Forth
Dunbar
56°N

North Sea

Alexandria
EAST DUNBARTONSHIRE
Falkirk
M9
Bathgate
Linlithgow
Edinburgh
EAST LOTHIAN
St Abb's Head
Dumbarton
Cumbernauld
Kirkintilloch
CITY OF EDINBURGH
Musselburgh
Eyemouth
Clydebank
Bearsden
Coatbridge
Airdrie
M8
Livingston
MIDLOTHIAN
LAMMERMUIR HILLS
Berwick-upon-Tweed
Paisley
Glasgow
NORTH LANARKSHIRE
WEST LOTHIAN
Penicuik
Duns
Whiteadder Water
Johnstone
GLASGOW CITY
PENTLAND HILLS
River Tweed
Coldstream
Holy Island
Hamilton
Motherwell
Wishaw
Bamburgh
RENFREWSHIRE
EAST RENFREWSHIRE
East Kilbride
M74
Lanark
Peebles
Innerleithen
Galashiels
Melrose
Kelso
Kilmarnock
Darvel
SOUTH LANARKSHIRE
Biggar
SCOTLAND
River Tweed
Broad Law 840m
Selkirk
SCOTTISH BORDERS
Wooler
EAST AYRSHIRE
River Ayr
Cumnock
Yarrow Water
St Mary's Loch
River Teviot
Jedburgh
815m The Cheviot
River Aln
Alnwick
Ayr
New Cumnock
LOWTHER HILLS
Daer Reservoir
Moffat
Hawick
CHEVIOT HILLS
602m Peel Fell
River Rede
NORTHUMBERLAND
River Coquet
Amble
Maybole
SOUTH AYRSHIRE
River Doon
Sanquhar
R. Annan
River Nith
Thornhill
Kielder Water
Ashington
Loch Doon
River Esk
Liddel Water
River Wansbeck
Blyth
DUMFRIES AND GALLOWAY
St John's Town of Dalry
Lochmaben
Lockerbie
Langholm
River Blyth
Whitley Bay
Cramlington
Newcastle upon Tyne
55°N
New Galloway
Loch Ken
Dumfries
River Annan
Annan
River Tyne
Gateshead
Washington
Newton Stewart
Castle Douglas
Dalbeattie
Kirkbean
Haltwhistle
River Irthing
Brampton
Hexham
River Derwent
Consett
Chester-le-Street
Wigtown
Gatehouse of Fleet
River Dee
Kirkcudbright
Solway Firth
Carlisle
Durham
Whithorn
Wigtown Bay
Wigton
M6
River Eden
Cross Fell 893m
River Wear
PENNINES
Spennymoor
Maryport
River Ellen
River Derwent
Cockermouth
Skiddaw 931m
Penrith
DURHAM
Bishop Auckland
Workington
Keswick
Derwent Water
CUMBRIA
Ullswater
Appleby-in-Westmorland
790m Mickle Fell
River Tees
Barnard Castle
Newton Aycliffe
DARLINGTON
Whitehaven
Helvellyn 950m
Brough
Darlington
St Bees Head
Scafell Pike 978m
LAKE DISTRICT
Kirkby Stephen
ENGLAND
Seascale
Ambleside
Windermere
Windermere
Richmond
NORTH YORKSHIRE
Coniston Water
Kendal
Whernside 737m
River Ure
Leyburn
Irish Sea
Point of Ayre
Snaefell 620m
Ramsey
Kirk Michael
ISLE OF MAN
South Barrule 483m
Douglas
Peel
Dalton-in-Furness
Barrow-in-Furness
Morecambe
Heysham
Carnforth
Morecambe Bay
Lancaster
M6
560m Ward's Stone
Ingleborough 723m
693m Pen-y-Ghent
704m Great Whernside
River Wharfe
R. Nidd
Ripon
54°N
Skipton

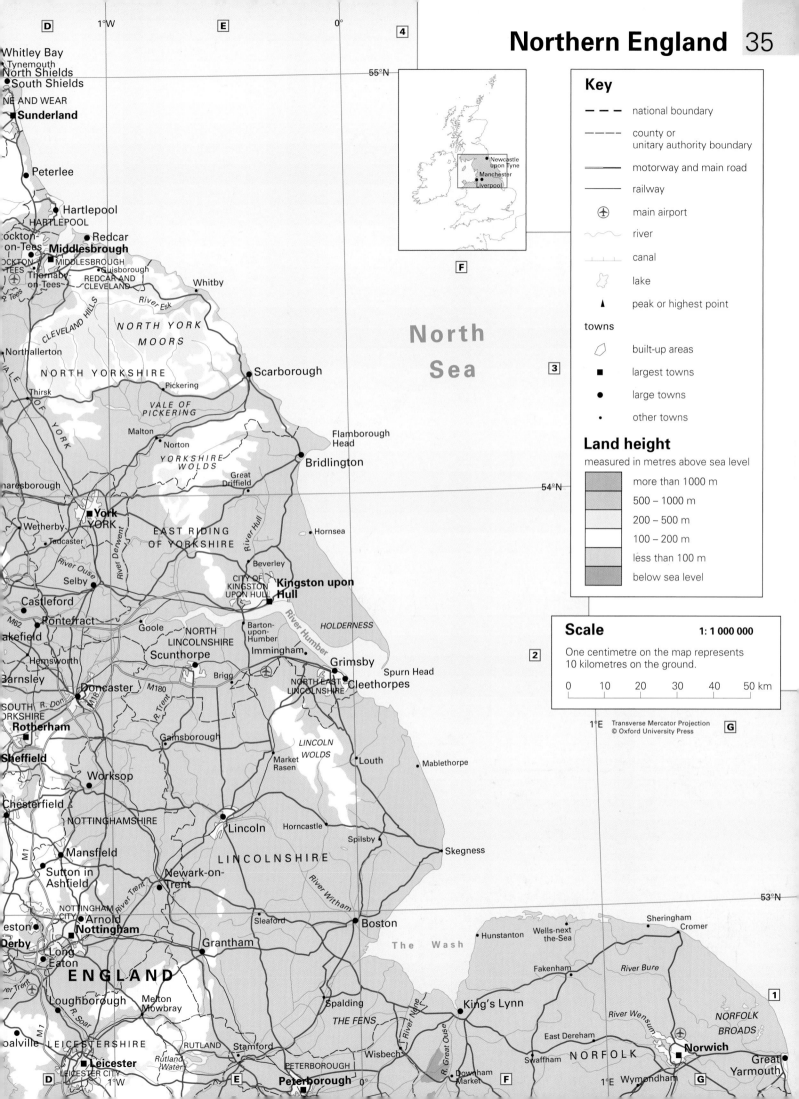

Key

- – – – national boundary
- – · – · – county or unitary authority boundary
- motorway and main road
- railway
- ✈ main airport
- river
- canal
- lake
- ▲ peak or highest point

towns
- built-up areas
- ■ largest towns
- ● large towns
- · other towns

Land height

measured in metres above sea level

- more than 1000 m
- 500 – 1000 m
- 200 – 500 m
- 100 – 200 m
- less than 100 m
- below sea level

Scale

1: 1 000 000

One centimetre on the map represents 10 kilometres on the ground.

| 0 | 10 | 20 | 30 | 40 | 50 km |

Transverse Mercator Projection
© Oxford University Press

North Sea

Whitley Bay
Tynemouth
North Shields
South Shields
NE AND WEAR
Sunderland
Peterlee
Hartlepool
HARTLEPOOL
ockton-on-Tees
Redcar
Middlesbrough
MIDDLESBROUGH
Thornaby-on-Tees
TEES
R. Tees
CLEVELAND HILLS
REDCAR AND CLEVELAND
Guisborough
Whitby
NORTH YORK MOORS
River Esk
Northallerton
NORTH YORKSHIRE
VALE OF YORK
Thirsk
Scarborough
Pickering
VALE OF PICKERING
Malton
Norton
YORKSHIRE WOLDS
Flamborough Head
Bridlington
Great Driffield
naresborough
54°N
York
YORK
Wetherby
Tadcaster
EAST RIDING OF YORKSHIRE
River Hull
Hornsea
River Derwent
River Ouse
Selby
Beverley
CITY OF KINGSTON UPON HULL
Kingston upon Hull
Castleford
Pontefract
Goole
HOLDERNESS
M62
akefield
NORTH LINCOLNSHIRE
Barton-upon-Humber
River Humber
Hemsworth
Scunthorpe
Immingham
Grimsby
Spurn Head
Barnsley
Brigg
NORTH EAST LINCOLNSHIRE
Cleethorpes
Doncaster
M180
M18
SOUTH YORKSHIRE
R. Don
R. Trent
Gainsborough
Rotherham
LINCOLN WOLDS
Louth
Mablethorpe
Sheffield
Worksop
Market Rasen
Chesterfield
NOTTINGHAMSHIRE
Lincoln
Horncastle
Spilsby
Skegness
M1
Mansfield
LINCOLNSHIRE
Sutton in Ashfield
Newark-on-Trent
River Trent
River Witham
53°N
NOTTINGHAM CITY
Arnold
Sleaford
Boston
Sheringham
Cromer
eston
Nottingham
ENGLAND
Grantham
The Wash
Hunstanton
Wells-next-the-Sea
Derby
Long Eaton
Fakenham
River Bure
r. Trent
Loughborough
R. Soar
Melton Mowbray
Spalding
King's Lynn
River Wensum
NORFOLK BROADS
oalville
LEICESTERSHIRE
RUTLAND
Stamford
THE FENS
River Nene
East Dereham
NORFOLK
Norwich
Great Yarmouth
Leicester
LEICESTER CITY
Rutland Water
PETERBOROUGH
Wisbech
R. Great Ouse
Swaffham
Wymondham
Peterborough
Downham Market

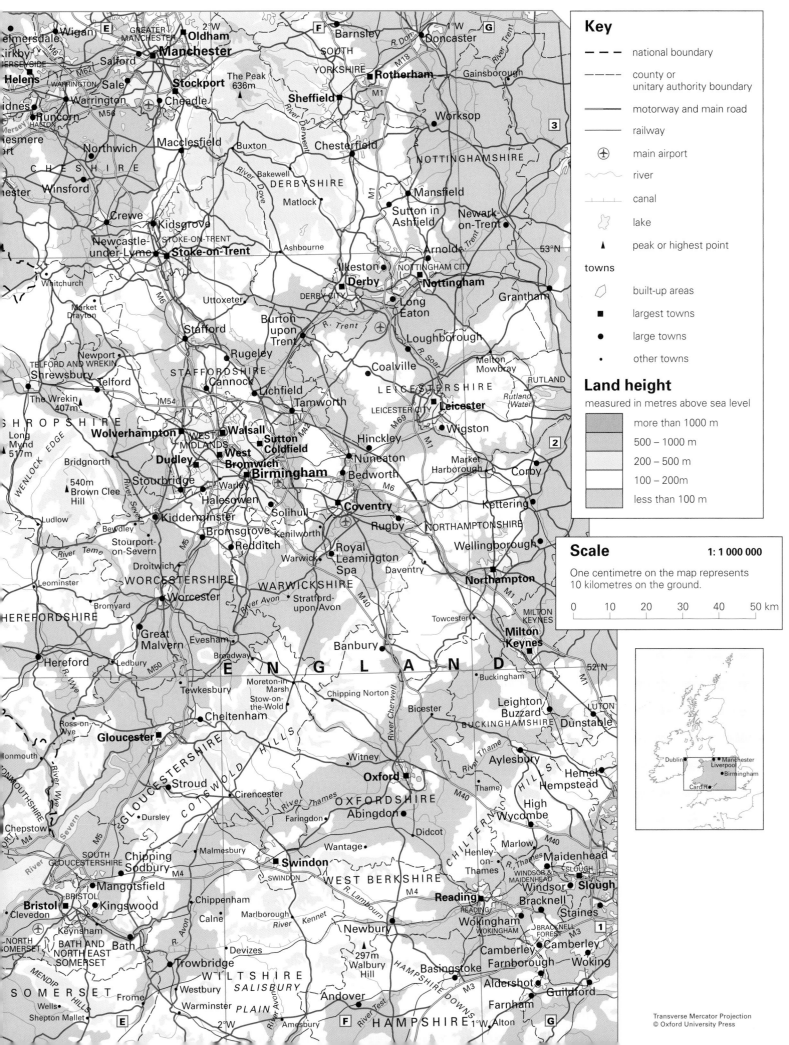

Key

- – – – national boundary
- – · – · – county or unitary authority boundary
- ——— motorway and main road
- ——— railway
- ⊕ main airport
- river
- canal
- lake
- ▲ peak or highest point

towns
- built-up areas
- ■ largest towns
- ● large towns
- · other towns

Land height

measured in metres above sea level

- more than 1000 m
- 500 – 1000 m
- 200 – 500 m
- 100 – 200m
- less than 100 m

Scale

1: 1 000 000

One centimetre on the map represents 10 kilometres on the ground.

0 10 20 30 40 50 km

Transverse Mercator Projection
© Oxford University Press

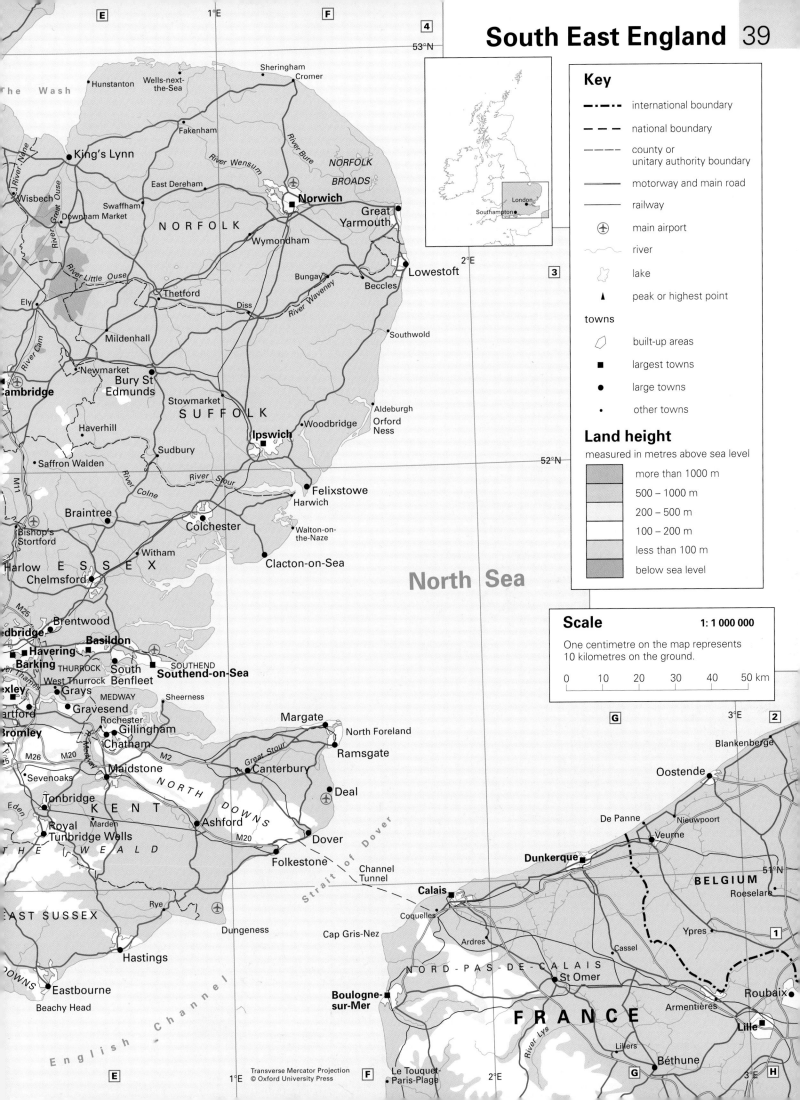

Key

- -·-·-· international boundary
- - - - national boundary
- -··-··- county or unitary authority boundary
- motorway and main road
- railway
- ✈ main airport
- river
- lake
- ▲ peak or highest point

towns

- built-up areas
- ■ largest towns
- ● large towns
- · other towns

Land height

measured in metres above sea level

- more than 1000 m
- 500 – 1000 m
- 200 – 500 m
- 100 – 200 m
- less than 100 m
- below sea level

Scale   1: 1 000 000

One centimetre on the map represents 10 kilometres on the ground.

0   10   20   30   40   50 km

North Sea

English Channel

Strait of Dover

Channel Tunnel

FRANCE

BELGIUM

NORD - PAS - DE - CALAIS

The Wash

NORFOLK BROADS

NORFOLK

SUFFOLK

ESSEX

KENT

NORTH DOWNS

THE WEALD

EAST SUSSEX

Transverse Mercator Projection
© Oxford University Press

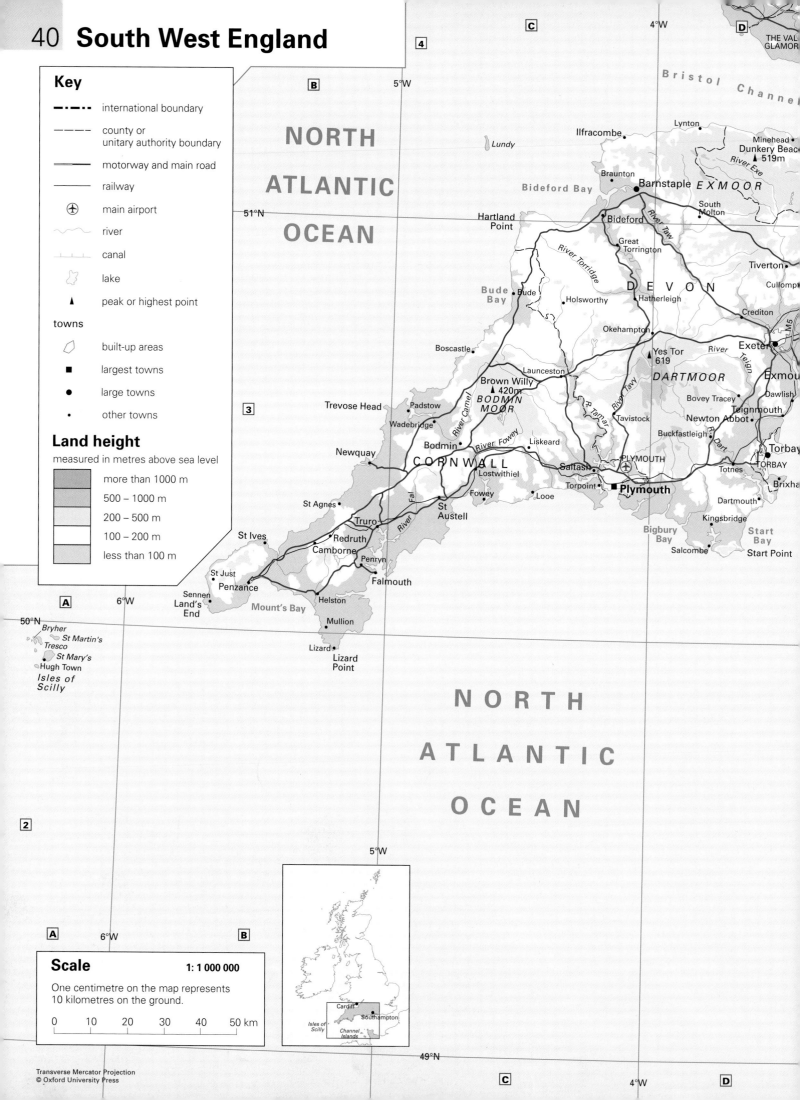

Key

▬·▬·▬	international boundary
▬ ▬ ▬	county or unitary authority boundary
▬▬▬	motorway and main road
▬▬▬	railway
⊕	main airport
∿	river
	canal
	lake
▲	peak or highest point

towns

⬠	built-up areas
■	largest towns
●	large towns
•	other towns

Land height

measured in metres above sea level

	more than 1000 m
	500 – 1000 m
	200 – 500 m
	100 – 200 m
	less than 100 m

NORTH ATLANTIC OCEAN

NORTH ATLANTIC OCEAN

Bristol Channel

THE VAL GLAMOR

Lundy
Ilfracombe
Lynton
Minehead
Dunkery Beac ▲ 519m
River Exe
Bideford Bay
Braunton
Barnstaple
South Molton
EXMOOR
Hartland Point
Bideford
Great Torrington
River Taw
Tiverton
Cullomp
DEVON
River Torridge
Bude Bay
Bude
Holsworthy
Hatherleigh
Crediton
M5
Okehampton
Boscastle
Yes Tor 619
Exeter
River Teign
DARTMOOR
Exmou
Launceston
Brown Willy ▲ 420m
BODMIN MOOR
River Tamar
Bovey Tracey
Dawlish
Trevose Head
Padstow
River Tavy
Newton Abbot
Teignmouth
Wadebridge
River Camel
R. Tavy
Tavistock
Buckfastleigh
R. Dart
Newquay
Bodmin
River Fowey
Liskeard
Torbay
St Agnes
CORNWALL
Lostwithiel
Saltash
PLYMOUTH
TORBAY
St Ives
Truro
Fal
St Austell
Fowey
Looe
Torpoint
Totnes
Brixha
Redruth
Camborne
River Fal
Plymouth
Dartmouth
St Just
Penryn
Kingsbridge
Penzance
Falmouth
Bigbury Bay
Start Bay
Sennen
Land's End
Helston
Salcombe
Start Point
Mount's Bay
Mullion
Lizard
Lizard Point

50°N
Bryher
St Martin's
Tresco
St Mary's
Hugh Town
Isles of Scilly

Scale

1: 1 000 000

One centimetre on the map represents 10 kilometres on the ground.

0 10 20 30 40 50 km

Cardiff
Southampton
Isles of Scilly
Channel Islands

D

M4

Cardiff 3°W
• Barry
Clevedon •
Bristol • **Kingswood**
BRISTOL
Keynsham
2°W Chippenham
Calne

E

M4

Weston-super-Mare
NORTH WEST SOMERSET
BATH AND NORTH EAST SOMERSET
Bath
• Devizes
Trowbridge
W I L T S H I R E
F
297m Walbury Hill
1°W Camberley Farnborough
Basingstoke
G
Woking Epsom
SURREY
Aldershot
Guildford
ridgwater Bay
MENDIP HILLS
Wells •
Westbury
Warminster
SALISBURY PLAIN
Andover
Dorking
Farnham
Alton
NORTH DOWNS

Shepton Mallet •
Frome
Amesbury
River Test
Stockbridge
R. Itchen
Haslemere
4

QUANTOCK HILLS
Bridgwater
Glastonbury •
Mere •
Salisbury
Winchester
H A M P S H I R E
Petersfield
Horsham
51°N

S O M E R S E T
Taunton
River Tone
River Parrett
Wincanton •
Shaftesbury
Romsey
Eastleigh
SOUTH DOWNS
WEST SUSSEX

Wellington •
M5
Ilchester •
River Yeo
Yeovil
Sherborne •
River Avon
Totton
Southampton
SOUTHAMPTON
Waterlooville
Havant
Arundel
Chichester
Worthing

Ilminster •
Crewkerne •
Chard •
Blandford Forum
Wimborne Minster
Ringwood •
Fawley
Fareham
Gosport
PORTSMOUTH
Portsmouth
Littlehampton
Bognor Regis

Honiton •
Axminster •
River Axe
Dorchester
River Frome
POOLE
Poole
BOURNEMOUTH
Bournemouth
Lymington •
Cowes
Newport
The Solent
Selsey Bill

Seaton •
Bridport •
Lyme Regis •
Wareham •
Christchurch
The Needles
• Ryde
ISLE OF WIGHT
Sandown
Shanklin

Sidmouth •
Lyme Bay
Weymouth
St Alban's Head
Swanage •
St Catherine's Point

Portland Bill

3

E n g l i s h   C h a n n e l

50°N

Cap de la Hague
Auderville •
Barfleur

Alderney
Cherbourg

Baie de la Seine

2

Guernsey
St Peter-Port
Sark •
Valognes

CHANNEL
F R A N C E

ISLANDS

Jersey
Carteret •
Carentan
Isigny-sur-Mer
Bayeux

St Helier •
Lessay •
River Vire
Caen

Coutainville •
Coutances •
St-Lô
49°N

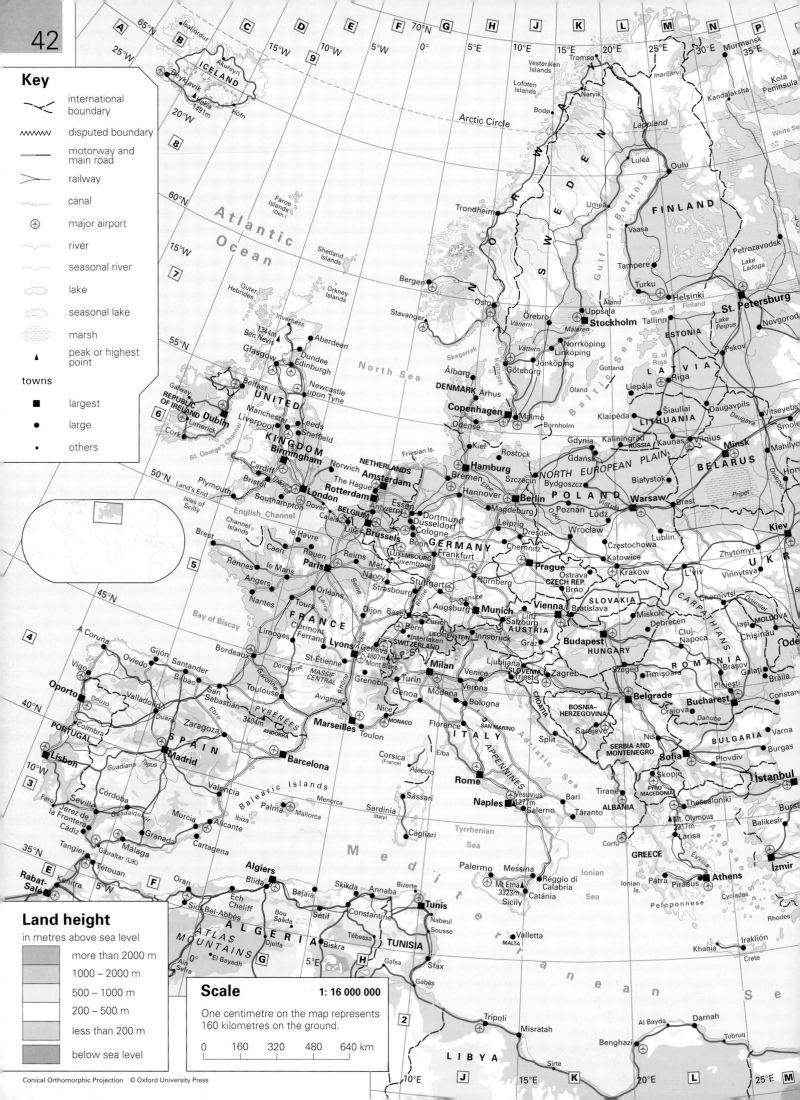

Key

- ⌇ international boundary
- ⩘ disputed boundary
- ⎯ motorway and main road
- ⤙ railway
- ⤙ canal
- ⊕ major airport
- river
- seasonal river
- lake
- seasonal lake
- marsh
- ▲ peak or highest point

towns

- ■ largest
- ● large
- · others

Land height

in metres above sea level

- more than 2000 m
- 1000 – 2000 m
- 500 – 1000 m
- 200 – 500 m
- less than 200 m
- below sea level

Scale

1: 16 000 000

One centimetre on the map represents 160 kilometres on the ground.

0   160   320   480   640 km

Conical Orthomorphic Projection   © Oxford University Press

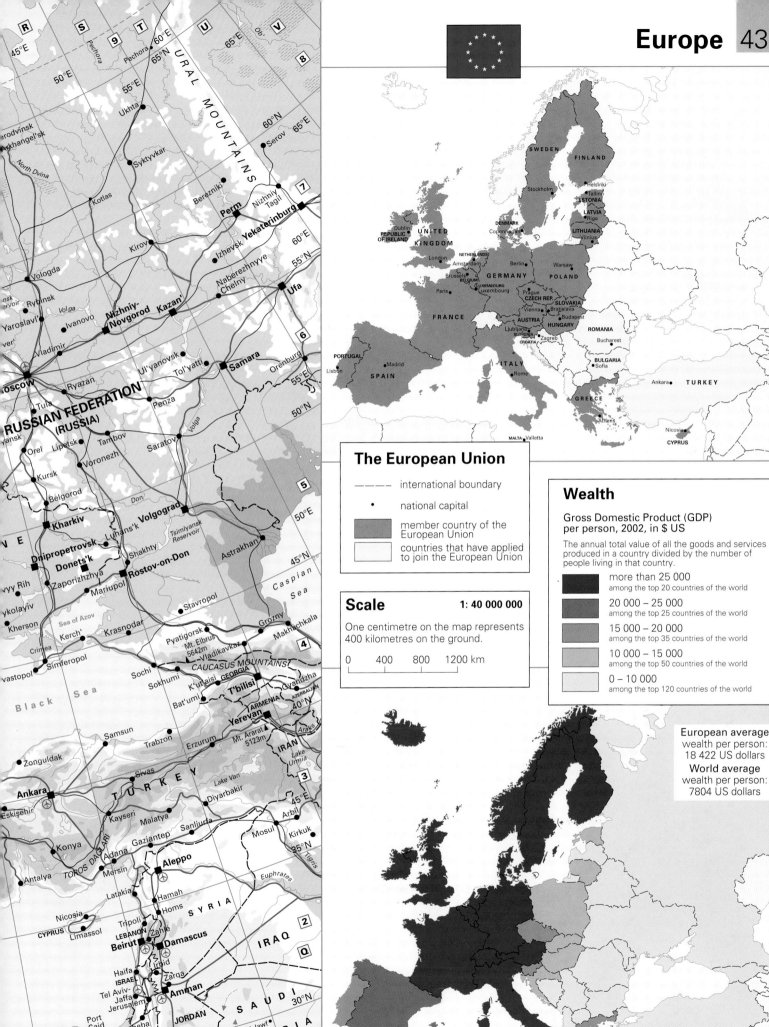

Map labels (left panel – physical/political map)

R | S | T | U | V
45°E | 50°E | 55°E · 60°E | 65°E | Ob'
Pechora | URAL MOUNTAINS | 9 | 8

Serodvinsk | Akhangel'sk | North Dvina
Ukhta | Serov | 60°N | 65°E
Syktyvkar | 7
Kotlas | Berezniki | Nizhniy Tagil
Perm | **Yekaterinburg** | 60°N
Kirov | Izhevsk | 60°E
Vologda | Naberezhnyye Chelny | 55°N
nsk | Rybinsk | Volga | Ivanovo | **Nizhniy-Novgorod** | **Kazan'** | **Ufa** | 6
Yaroslavl' | Vladimir | Ul'yanovsk | Tol'yatti | **Samara** | Orenburg | 55°E
ver' | Moscow | Ryazan | Penza | Volga | 50°N
Tula | **RUSSIAN FEDERATION (RUSSIA)** | Saratov
Orel | Lipetsk | Tambov | 5
yansk | Voronezh | Don
Kursk | Belgorod | Tsimlyansk Reservoir
Kharkiv | **Volgograd** | 50°E
vyy Rih | **Dnipropetrovsk** | Luhans'k | Shakhty | Astrakhan | 45°N
ykolayiv | **Donets'k** | Zaporizhzhya | **Rostov-on-Don** | Caspian Sea
Kherson | Mariupol | Stavropol
Kerch' | Krasnodar | Grozny | Makhachkala | 4
vastopol | Crimea | Simferopol | Pyatigorsk | Mt. Elbrus 5642m | Vladikavkaz | Grozny
Sochi | CAUCASUS MOUNTAINS | Sokhumi | K'ut'aisi GEORGIA | Gyandzha
Black Sea | Bat'umi | **T'bilisi** | AZERBAIJAN | 40°N
Samsun | ARMENIA | **Yerevan** | Araks
Zonguldak | Trabzon | Erzurum | Mt. Ararat 5123m | IRAN | Lake Urmia | 3
Sivas | Lake Van | 45°E
Ankara | TURKEY | Diyarbakir | 40°N
Eskişehir | Kayseri | Malatya | Sanliurfa | Arbil
Konya | Adana | Gaziantep | Mosul | Kirkuk | 35°N
Antalya | TOROS DAĞLARI | Mersin | **Aleppo** | Tigris
Latakia | Hamah | Euphrates | 2
Nicosia | Homs | SYRIA | IRAQ
CYPRUS | Limassol | Tripoli | Zahle | LEBANON
Beirut | **Damascus** | Q
Haifa | Irbid | Zarqa
ISRAEL | Tel Aviv-Jaffa | **Amman** | 30°N
Jerusalem | JORDAN | SAUDI ARABIA | 1
Port Said | Al Jawf
xandria | Beersheba | Aqaba
EGYPT | Giza **Cairo** | Suez | Tabuk
30°E | N | 35°E | P | 40°E

The European Union

- – – – international boundary
- • national capital
- ▨ member country of the European Union
- ▢ countries that have applied to join the European Union

Scale

1 : 40 000 000

One centimetre on the map represents 400 kilometres on the ground.

0 — 400 — 800 — 1200 km

Wealth

Gross Domestic Product (GDP) per person, 2002, in $ US

The annual total value of all the goods and services produced in a country divided by the number of people living in that country.

- more than 25 000 — among the top 20 countries of the world
- 20 000 – 25 000 — among the top 25 countries of the world
- 15 000 – 20 000 — among the top 35 countries of the world
- 10 000 – 15 000 — among the top 50 countries of the world
- 0 – 10 000 — among the top 120 countries of the world

European average wealth per person: 18 422 US dollars

World average wealth per person: 7804 US dollars

EU member map labels

SWEDEN | FINLAND
Stockholm | Helsinki | Tallinn | ESTONIA
Dublin | DENMARK | LATVIA | Riga
REPUBLIC OF IRELAND | UNITED KINGDOM | Copenhagen | LITHUANIA | Vilnius
London | NETHERLANDS | Amsterdam | Berlin | Warsaw
Brussels | BELGIUM | GERMANY | POLAND
Paris | LUXEMBOURG | Luxembourg | Prague | CZECH REP.
FRANCE | SLOVAKIA | Bratislava | Budapest
AUSTRIA | Vienna | HUNGARY | ROMANIA
Ljubljana | SLOVENIA | Zagreb | Bucharest
PORTUGAL | CROATIA | BULGARIA | Sofia
Lisbon | SPAIN | Madrid | ITALY | Rome | Ankara | TURKEY
GREECE | Athens | Nicosia
MALTA | Valletta | CYPRUS

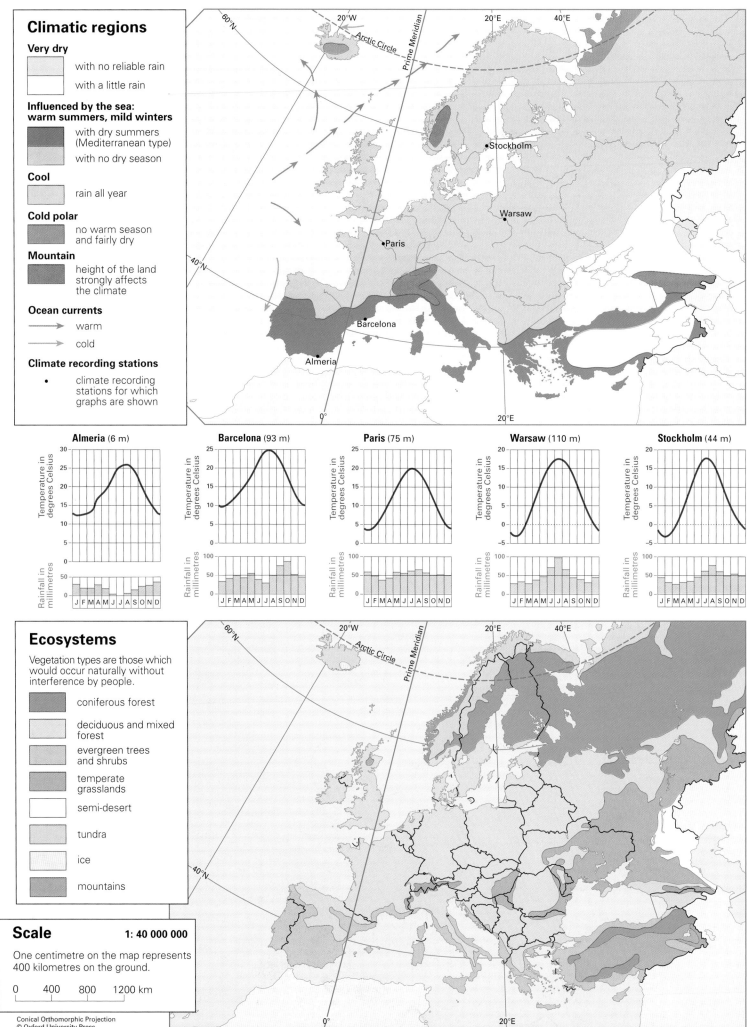

Climatic regions

Very dry
- with no reliable rain
- with a little rain

Influenced by the sea: warm summers, mild winters
- with dry summers (Mediterranean type)
- with no dry season

Cool
- rain all year

Cold polar
- no warm season and fairly dry

Mountain
- height of the land strongly affects the climate

Ocean currents
- → warm
- → cold

Climate recording stations
- • climate recording stations for which graphs are shown

Almeria (6 m)

Temperature in degrees Celsius / Rainfall in millimetres
J F M A M J J A S O N D

Barcelona (93 m)

Temperature in degrees Celsius / Rainfall in millimetres
J F M A M J J A S O N D

Paris (75 m)

Temperature in degrees Celsius / Rainfall in millimetres
J F M A M J J A S O N D

Warsaw (110 m)

Temperature in degrees Celsius / Rainfall in millimetres
J F M A M J J A S O N D

Stockholm (44 m)

Temperature in degrees Celsius / Rainfall in millimetres
J F M A M J J A S O N D

Ecosystems

Vegetation types are those which would occur naturally without interference by people.

- coniferous forest
- deciduous and mixed forest
- evergreen trees and shrubs
- temperate grasslands
- semi-desert
- tundra
- ice
- mountains

Scale

1: 40 000 000

One centimetre on the map represents 400 kilometres on the ground.

0   400   800   1200 km

Conical Orthomorphic Projection
© Oxford University Press

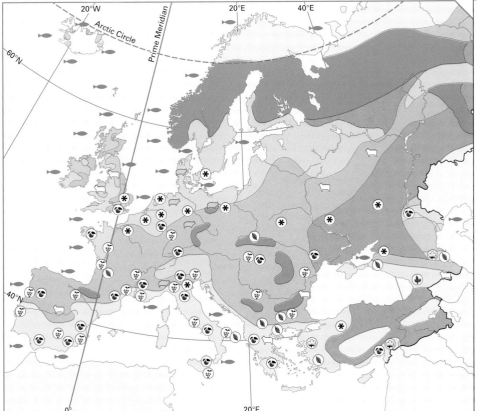

Farming, forestry, and fishing

main farming types

	little or no farming : because the area is too cold or otherwise harsh.
	nomadic herding : animals provide all the needs of the wandering families.
	shifting cultivation : small areas farmed until soils exhausted, then family moves.
	mixed subsistence : crops and animals for family food.
	grazing and stock rearing : on a large scale, for profit.
	mixed farming : animals and crops for profit.
	grain farming : mostly wheat, on a large scale, for profit.
	mediterranean farming : cereals, animals, vegetables, fruit, wine sold for profit.
	specialized horticulture : often supported by irrigation.
	dairy farming : milk, butter, and cheese for profit.

forestry

	cutting and replacement of timber for profit

cash crops

🍇	wine grapes	🍵	tea	🚬	tobacco
🍑	fruit	✳	sugar		cotton

animal products

🐑	wool	🐖	meat	🐟	fish

Almeria
Mean annual rainfall: 233 mm   Mean January temperature: 12.0°C   Mean July temperature: 25.0°C

Barcelona
Mean annual rainfall: 587 mm   Mean January temperature: 9.5°C   Mean July temperature: 24.5°C

Paris
Mean annual rainfall: 589 mm   Mean January temperature: 3.5°C   Mean July temperature: 20.0°C

Warsaw
Mean annual rainfall: 525 mm   Mean January temperature: -3.0°C   Mean July temperature: 19.5°C

Stockholm
Mean annual rainfall: 524 mm   Mean January temperature: -3.0°C   Mean July temperature: 18.0°C

Scale

1 : 40 000 000

One centimetre on the map represents 400 kilometres on the ground.

0   400   800   1200 km

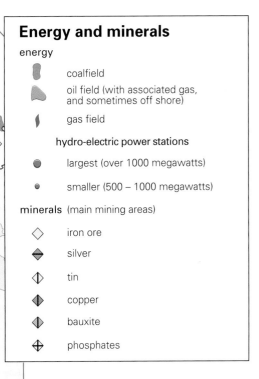

Energy and minerals

energy

	coalfield
	oil field (with associated gas, and sometimes off shore)
	gas field

hydro-electric power stations

●	largest (over 1000 megawatts)
•	smaller (500 – 1000 megawatts)

minerals (main mining areas)

◇	iron ore
◈	silver
◁▷	tin
◈	copper
◁▷	bauxite
✣	phosphates

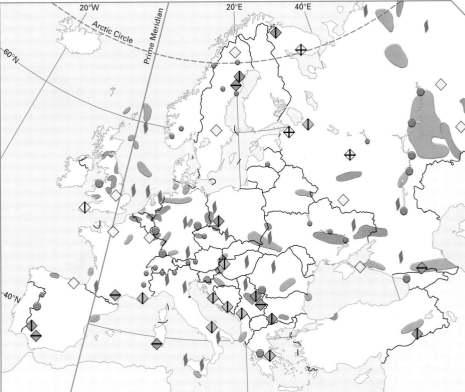

Conical Orthomorphic Projection
© Oxford University Press

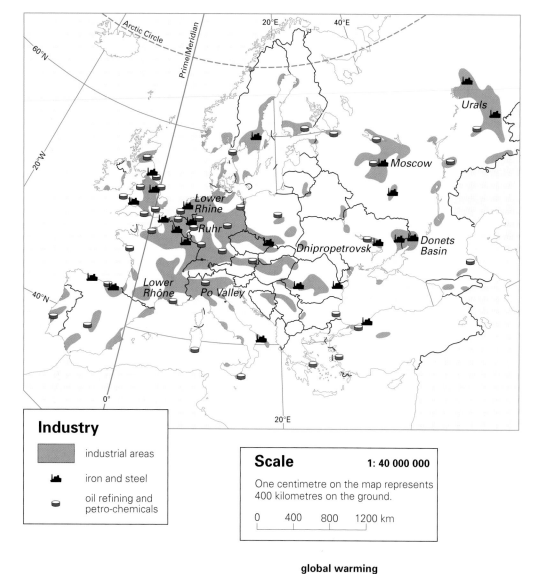

Industry

▨	industrial areas
⚒	iron and steel
⬭	oil refining and petro-chemicals

Scale
1: 40 000 000

One centimetre on the map represents 400 kilometres on the ground.

0	400	800	1200 km

Population structure of the United Kingdom

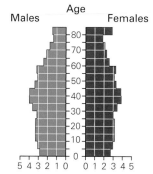

Age
Males | Females

5 4 3 2 1 0 0 1 2 3 4 5

percent of total population in 2004
Total population : 60.3 million

Population structure of France

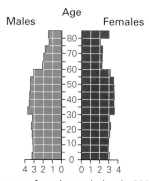

Age
Males | Females

4 3 2 1 0 0 1 2 3 4

percent of total population in 2004
Total population : 60.4 million

Environmental issues

sea pollution

■	areas severely polluted for all or part of the year
▨	areas persistently affected by pollution
▼	deep sea dump sites
✳	major oil spills (over 100 000 tonnes)
∗	major oil spills (under 100 000 tonnes)

acid rain

A pH scale measures acidity. Unaffected rain water is slightly acidic with a pH of 5.6

▨	pH less than 4.2 (most acidic)
▨	pH 4.2 – 4.6
▨	pH 4.6 – 5.0

air pollution

◆	cities where sulphur dioxide emissions are recorded and exceed recommended levels

industrial sites emitting the largest amounts of sulphur

◯	over 200 000 tonnes
◯	100 000 – 200 000 tonnes
○	50 000 – 100 000 tonnes
○	30 000 – 50 000 tonnes

global warming
addition of greenhouse gases in tonnes of carbon per person
(look at the world map on page 17)

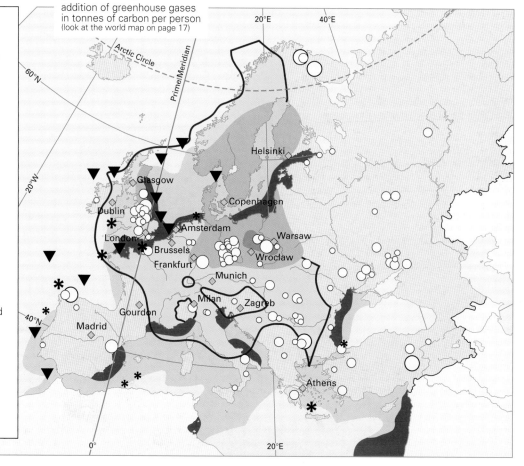

Conical Orthomorphic Projection
© Oxford University Press

Population structure of Germany

Males — Age — Females

```
        80
        70
        60
        50
        40
        30
        20
        10
5 4 3 2 1 0   0 1 2 3 4 5
```

percent of total population in 2004
Total population : 82.4 million

Population structure of Greece

Males — Age — Females

```
        80
        70
        60
        50
        40
        30
        20
        10
         0
4 3 2 1 0   0 1 2 3 4
```

percent of total population in 2004
Total population : 10.6 million

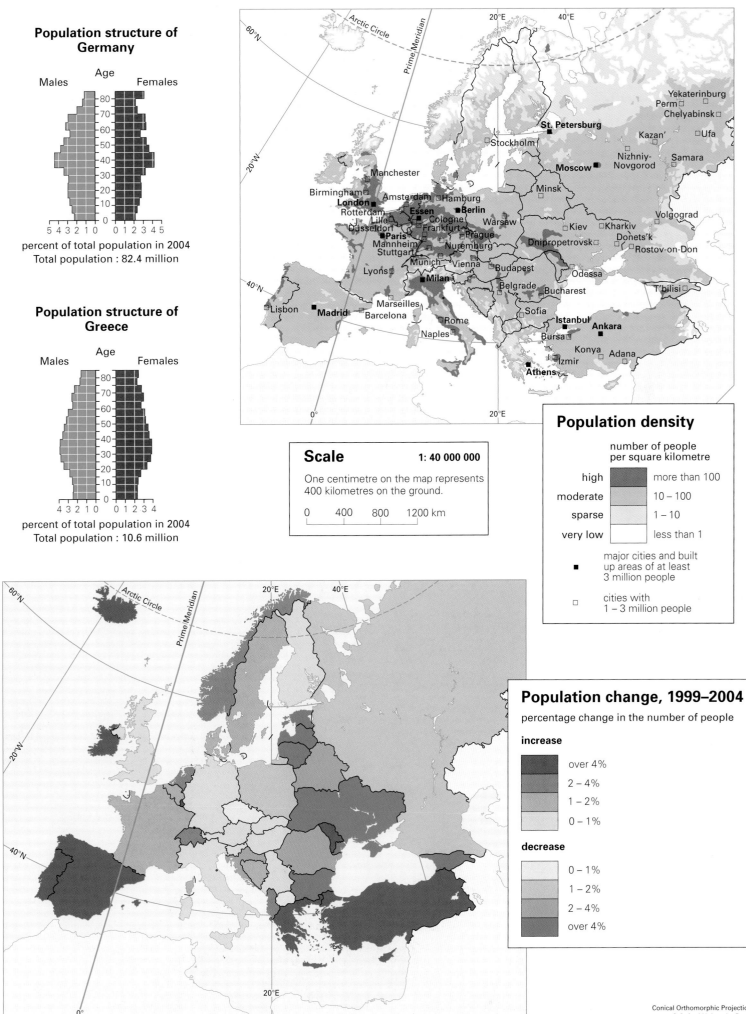

Scale

1 : 40 000 000

One centimetre on the map represents 400 kilometres on the ground.

```
0    400   800   1200 km
```

Population density

number of people
per square kilometre

high	more than 100
moderate	10 – 100
sparse	1 – 10
very low	less than 1

■ major cities and built up areas of at least 3 million people

□ cities with 1 – 3 million people

Population change, 1999–2004

percentage change in the number of people

increase

	over 4%
	2 – 4%
	1 – 2%
	0 – 1%

decrease

	0 – 1%
	1 – 2%
	2 – 4%
	over 4%

Conical Orthomorphic Projection
© Oxford University Press

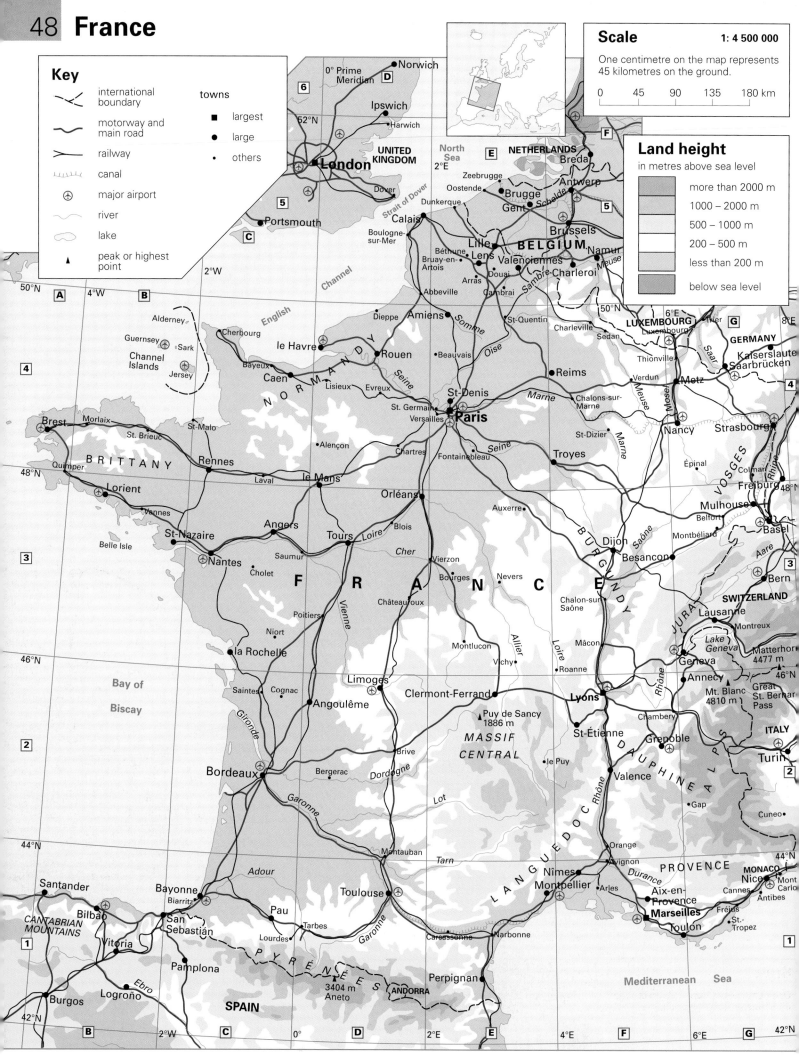

Scale 1: 4 500 000

One centimetre on the map represents 45 kilometres on the ground.

0   45   90   135   180 km

Key

- international boundary
- motorway and main road
- railway
- canal
- major airport
- river
- lake
- peak or highest point

towns
- ■ largest
- ● large
- • others

Land height
in metres above sea level

- more than 2000 m
- 1000 – 2000 m
- 500 – 1000 m
- 200 – 500 m
- less than 200 m
- below sea level

Conical Orthomorphic Projection   © Oxford University Pr

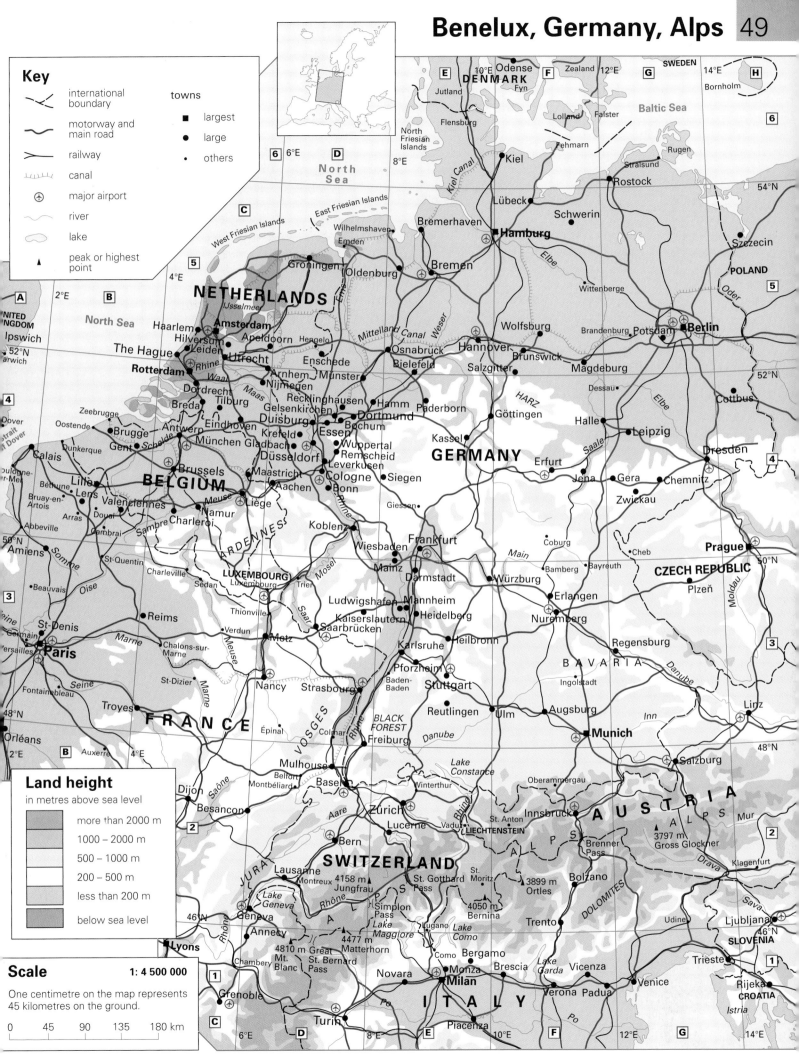

Key

	international boundary
	motorway and main road
	railway
	canal
	major airport
	river
	lake
	peak or highest point

towns

- ■ largest
- ● large
- • others

Land height

in metres above sea level

- more than 2000 m
- 1000 – 2000 m
- 500 – 1000 m
- 200 – 500 m
- less than 200 m
- below sea level

Scale

1: 4 500 000

One centimetre on the map represents 45 kilometres on the ground.

0   45   90   135   180 km

nical Orthomorphic Projection   © Oxford University Press

DENMARK
SWEDEN
Baltic Sea
Jutland
Flensburg
North Friesian Islands
Fehmarn
Lolland
Falster
Zealand
Fyn
Odense
Bornholm
Rugen
Stralsund
Kiel
Kiel Canal
Rostock
Lübeck
Schwerin
POLAND
Szezecin
Bremerhaven
Hamburg
Elbe
Wilhelmshaven
Emden
Oldenburg
Bremen
Wittenberge
Oder
East Friesian Islands
Groningen
Ems
Wolfsburg
Brandenburg
Potsdam
Berlin
NETHERLANDS
West Friesian Islands
IJsselmeer
Mittelland Canal
Weser
Osnabrück
Hannover
Brunswick
Magdeburg
Dessau
Cottbus
North Sea
Haarlem
Amsterdam
Hilversum
Apeldoorn
Hengelo
Enschede
Bielefeld
Salzgitter
Elbe
Halle
Leipzig
Dresden
The Hague
Leiden
Utrecht
Arnhem
Münster
Paderborn
Göttingen
Kassel
HARZ
Saale
Rhine
Rotterdam
Waal
Nijmegen
Recklinghausen
Hamm
GERMANY
Erfurt
Jena
Gera
Chemnitz
Dordrecht
Maas
Tilburg
Gelsenkirchen
Dortmund
Bochum
Breda
Eindhoven
Duisburg
Essen
Wuppertal
Remscheid
Siegen
Zwickau
Zeebrugge
Krefeld
München Gladbach
Düsseldorf
Leverkusen
Giessen
Oostende
Brugge
Antwerp
Schelde
Maastricht
Aachen
Cologne
Bonn
Dresden
Calais
Dunkerque
Gent
BELGIUM
Brussels
Liège
Koblenz
Rhine
Coburg
Cheb
Prague
Dover
Strait of Dover
Lille
Lens
Valenciennes
Meuse
Namur
Charleroi
ARDENNES
Frankfurt
Main
Bamberg
Bayreuth
CZECH REPUBLIC
Béthune
Bruay-en-Artois
Douai
Sambre
Wiesbaden
Mainz
Würzburg
Plzeň
Moldau
Arras
Cambrai
Charleville
LUXEMBOURG
Mosel
Darmstadt
Erlangen
Abbeville
St-Quentin
Sédan
Luxembourg
Trier
Ludwigshafen
Mannheim
Nuremberg
Regensburg
Linz
Amiens
Somme
Thionville
Saar
Kaiserslautern
Heidelberg
Danube
BAVARIA
Beauvais
Oise
Verdun
Metz
Saarbrücken
Karlsruhe
Heilbronn
Ingolstadt
St-Denis
Reims
Marne
Meuse
Pforzheim
Stuttgart
Augsburg
Seine
Paris
Chalons-sur-Marne
Nancy
Strasbourg
Baden-Baden
Reutlingen
Ulm
Munich
Salzburg
Versailles
St-Germain
St-Dizier
Marne
VOSGES
Colmar
Rhine
BLACK FOREST
Danube
Troyes
Épinal
Freiburg
Lake Constance
Oberammergau
AUSTRIA
Fontainebleau
Seine
FRANCE
Mulhouse
Winterthur
Innsbruck
ALPS
Mur
Orléans
Auxerre
Belfort
Montbéliard
Basel
Zürich
St. Anton
Vaduz
LIECHTENSTEIN
Brenner Pass
3797 m Gross Glockner
Klagenfurt
Dijon
Besançon
Aare
Lucerne
Rhine
ALPS
Drava
JURA
Bern
St. Gotthard Pass
St. Moritz
3899 m Ortles
Bolzano
SWITZERLAND
Lausanne
Montreux
4158 m Jungfrau
4050 m Bernina
Trento
DOLOMITES
Ljubljana
Sava
Saône
Lake Geneva
Rhône
Simplon Pass
Lake Maggiore
Lugano
Lake Como
SLOVENIA
46°N
Geneva
Annecy
A L P S
4477 m Matterhorn
Como
Bergamo
Brescia
Lake Garda
Vicenza
Venice
Udine
Trieste
Istria
Chambery
4810 m Mt. Blanc
Great St. Bernard Pass
Monza
Brescia
Verona
Padua
Rijeka
CROATIA
Grenoble
Milan
ITALY
Po
Turin
Piacenza

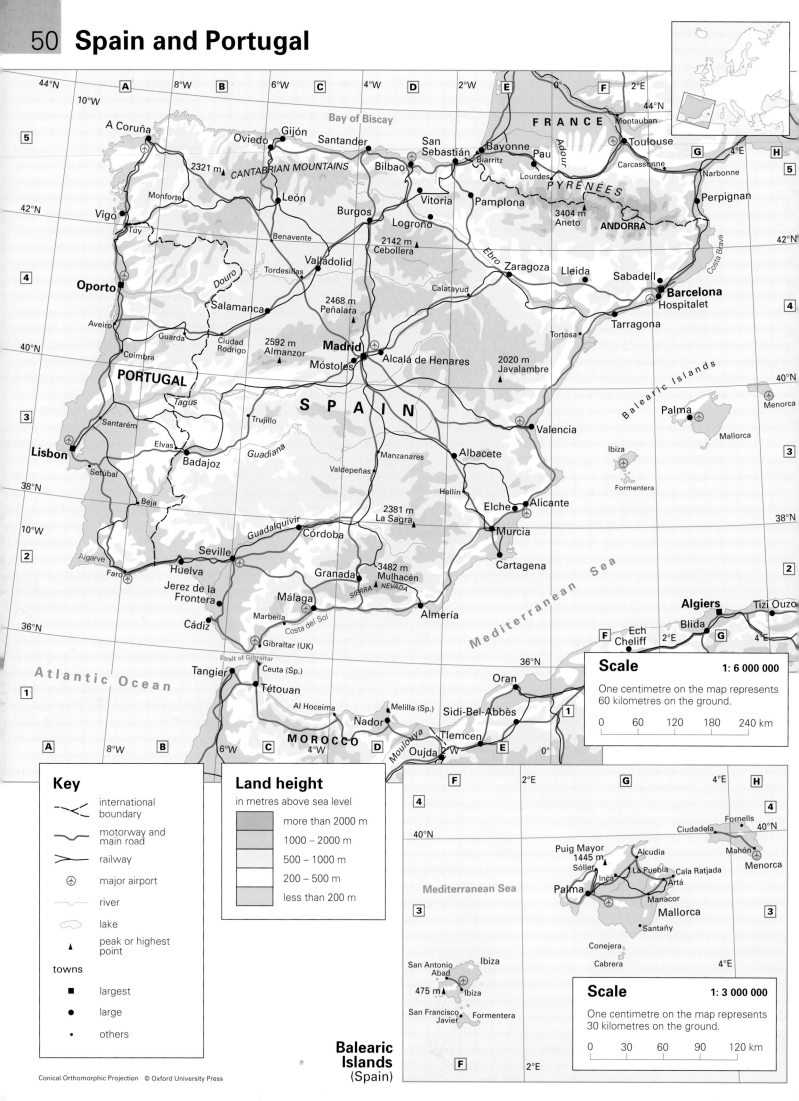

Key

- –··–‹ international boundary
- ⌒ motorway and main road
- ⤚⤙ railway
- ⊕ major airport
- ⌇ river
- ⬡ lake
- ▲ peak or highest point

towns

- ■ largest
- ● large
- · others

Land height

in metres above sea level

- more than 2000 m
- 1000 – 2000 m
- 500 – 1000 m
- 200 – 500 m
- less than 200 m

Scale 1: 6 000 000

One centimetre on the map represents 60 kilometres on the ground.

0 60 120 180 240 km

Scale 1: 3 000 000

One centimetre on the map represents 30 kilometres on the ground.

0 30 60 90 120 km

Balearic Islands (Spain)

Conical Orthomorphic Projection © Oxford University Press

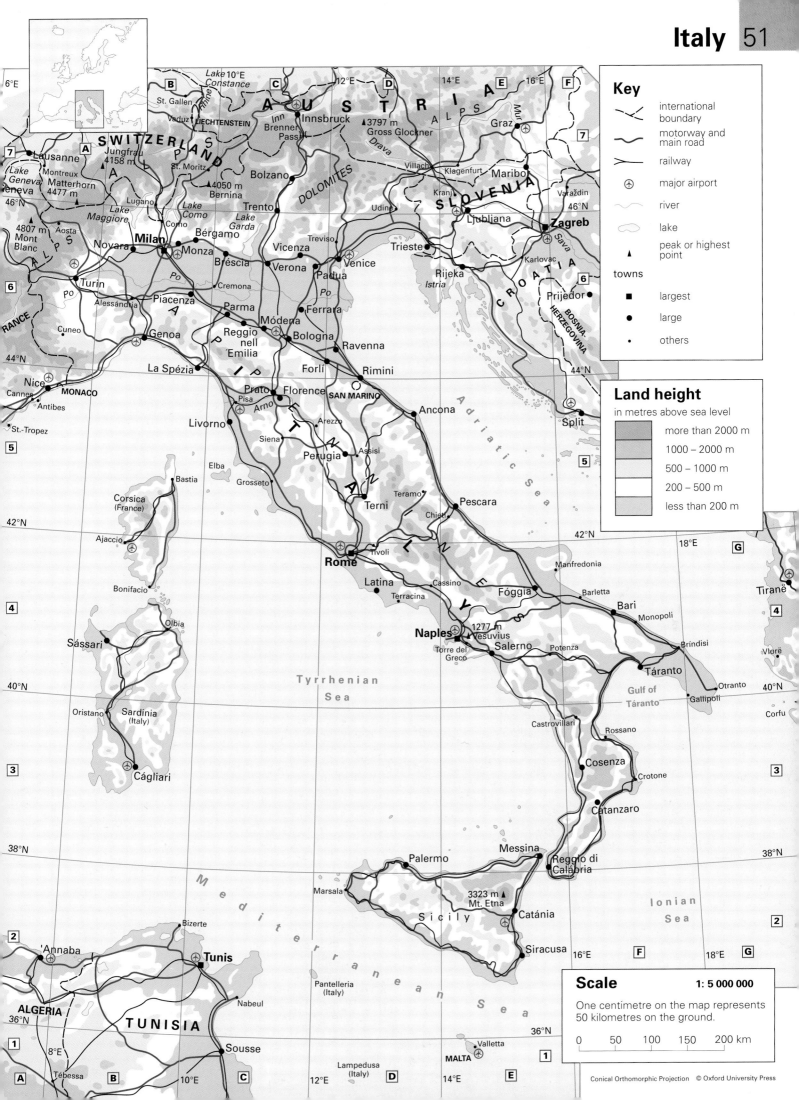

Key

- ‒‒ international boundary
- ‒‒ motorway and main road
- ‒‒ railway
- ⊕ major airport
- ～ river
- ◯ lake
- ▲ peak or highest point

towns
- ■ largest
- ● large
- • others

Land height
in metres above sea level

- more than 2000 m
- 1000 – 2000 m
- 500 – 1000 m
- 200 – 500 m
- less than 200 m

Scale
1: 5 000 000

One centimetre on the map represents 50 kilometres on the ground.

0   50   100   150   200 km

Conical Orthomorphic Projection   © Oxford University Press

SWITZERLAND
St. Gallen
Vaduz LIECHTENSTEIN
Lake Constance
Inn
Innsbruck
Brenner Pass
AUSTRIA
3797 m Gross Glockner
Drava
Graz
Villach
Klagenfurt
Maribo
Lausanne
Montreux
Jungfrau 4158 m
Matterhorn 4477 m
St. Moritz
Bernina 4050 m
Bolzano
DOLOMITES
Trento
Udine
SLOVENIA
Kranj
Ljubljana
Varaždin
Zagreb
Lake Geneva Geneva
Mont Blanc 4807 m
Aosta
Lugano
Lake Maggiore
Lake Como
Como
Lake Garda
Bérgamo
Bréscia
Verona
Vicenza
Treviso
Venice
Trieste
Rijeka
Istria
Karlovac
CROATIA
Prijedor
FRANCE
Cuneo
Novara
Milan
Monza
Turin
Po
Alessándria
Piacenza
Cremona
Parma
Po
Módena
Reggio nell'Emilia
Bologna
Ferrara
Ravenna
Rimini
Forlí
BOSNIA HERZEGOVINA
Nice
Cannes
Antibes
St.-Tropez
MONACO
Genoa
La Spézia
Prato
Pisa
Arno
Florence
SAN MARINO
Ancona
Adriatic Sea
Split
Livorno
Elba
Arezzo
Siena
Grosseto
Perugia
Assisi
APENNINES
Teramo
Terni
Chieti
Pescara
Bastia
Corsica (France)
Ajaccio
Bonifacio
Tivoli
Rome
Latina
Terracina
Cassino
Manfredonia
Barletta
Bari
Monopoli
Bríndisi
Tiranë
Vlorë
Olbia
Sássari
Oristano
Sardinia (Italy)
Cágliari
Naples
Vesuvius 1277 m
Torre del Greco
Salerno
Potenza
Fóggia
Táranto
Gulf of Táranto
Otranto
Gallipoli
Corfu
Tyrrhenian Sea
Castrovillari
Rossano
Cosenza
Crotone
Catanzaro
'Annaba
Tunis
ALGERIA
TUNISIA
Bizerte
Nabeul
Sousse
Pantelleria (Italy)
Lampedusa (Italy)
Marsala
Palermo
Messina
Reggio di Calábria
Mt. Etna 3323 m
Catánia
Siracusa
Sicily
Mediterranean Sea
Ionian Sea
Valletta
MALTA

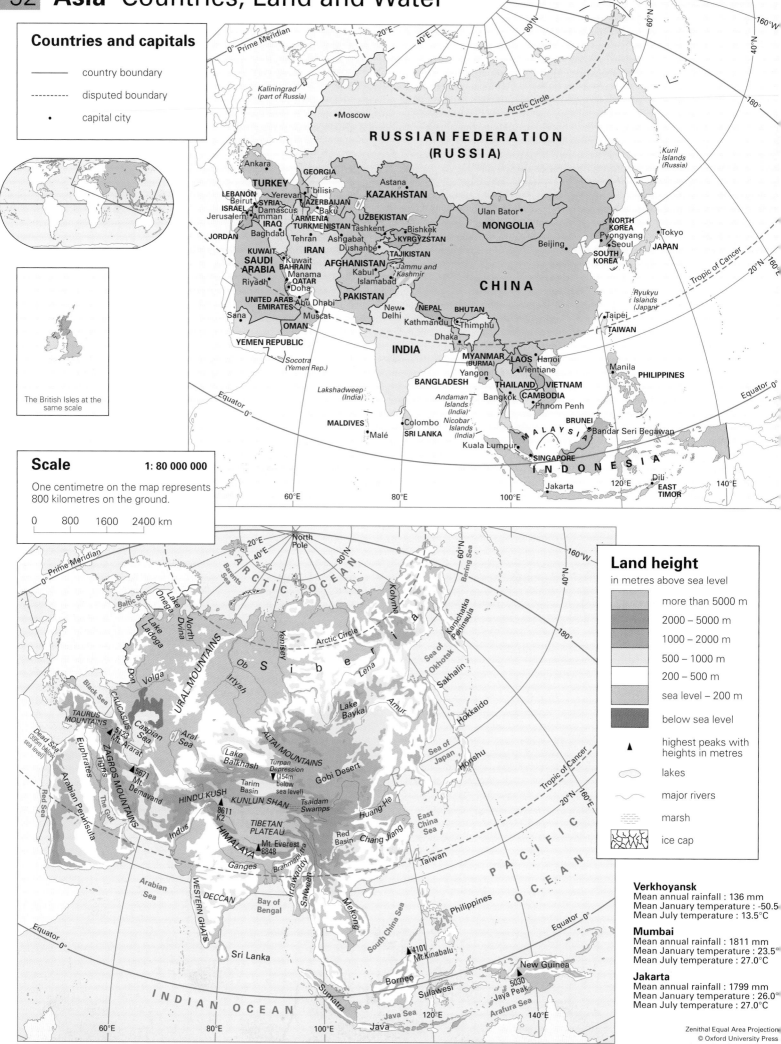

Countries and capitals

——	country boundary
- - -	disputed boundary
•	capital city

The British Isles at the same scale

Scale
1: 80 000 000

One centimetre on the map represents 800 kilometres on the ground.

0   800   1600   2400 km

Land height
in metres above sea level

- more than 5000 m
- 2000 – 5000 m
- 1000 – 2000 m
- 500 – 1000 m
- 200 – 500 m
- sea level – 200 m
- below sea level
- ▲ highest peaks with heights in metres
- lakes
- major rivers
- marsh
- ice cap

Verkhoyansk
Mean annual rainfall : 136 mm
Mean January temperature : -50.5
Mean July temperature : 13.5°C

Mumbai
Mean annual rainfall : 1811 mm
Mean January temperature : 23.5°C
Mean July temperature : 27.0°C

Jakarta
Mean annual rainfall : 1799 mm
Mean January temperature : 26.0°C
Mean July temperature : 27.0°C

Zenithal Equal Area Projection
© Oxford University Press

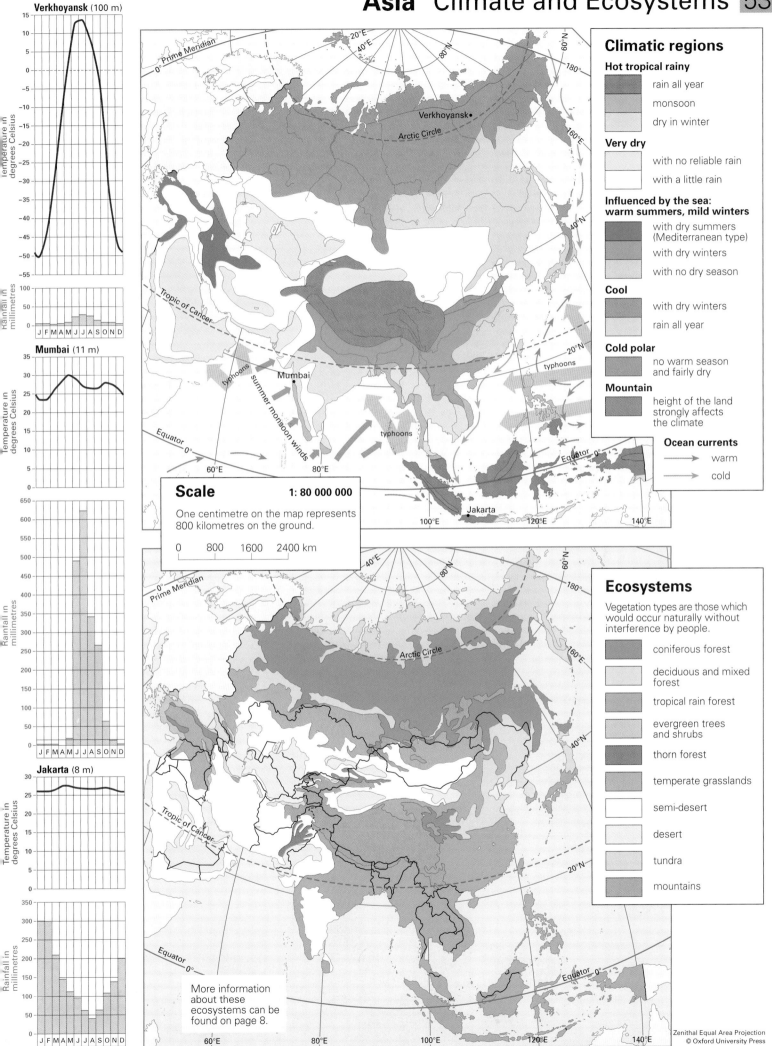

Verkhoyansk (100 m)

Temperature in degrees Celsius

Rainfall in millimetres

JFMAMJJASOND

Mumbai (11 m)

Temperature in degrees Celsius

Rainfall in millimetres

JFMAMJJASOND

Jakarta (8 m)

Temperature in degrees Celsius

Rainfall in millimetres

JFMAMJJASOND

Climatic regions

Hot tropical rainy
- rain all year
- monsoon
- dry in winter

Very dry
- with no reliable rain
- with a little rain

Influenced by the sea: warm summers, mild winters
- with dry summers (Mediterranean type)
- with dry winters
- with no dry season

Cool
- with dry winters
- rain all year

Cold polar
- no warm season and fairly dry

Mountain
- height of the land strongly affects the climate

Ocean currents
→ warm
→ cold

Scale

1: 80 000 000

One centimetre on the map represents 800 kilometres on the ground.

0   800   1600   2400 km

Ecosystems

Vegetation types are those which would occur naturally without interference by people.

- coniferous forest
- deciduous and mixed forest
- tropical rain forest
- evergreen trees and shrubs
- thorn forest
- temperate grasslands
- semi-desert
- desert
- tundra
- mountains

More information about these ecosystems can be found on page 8.

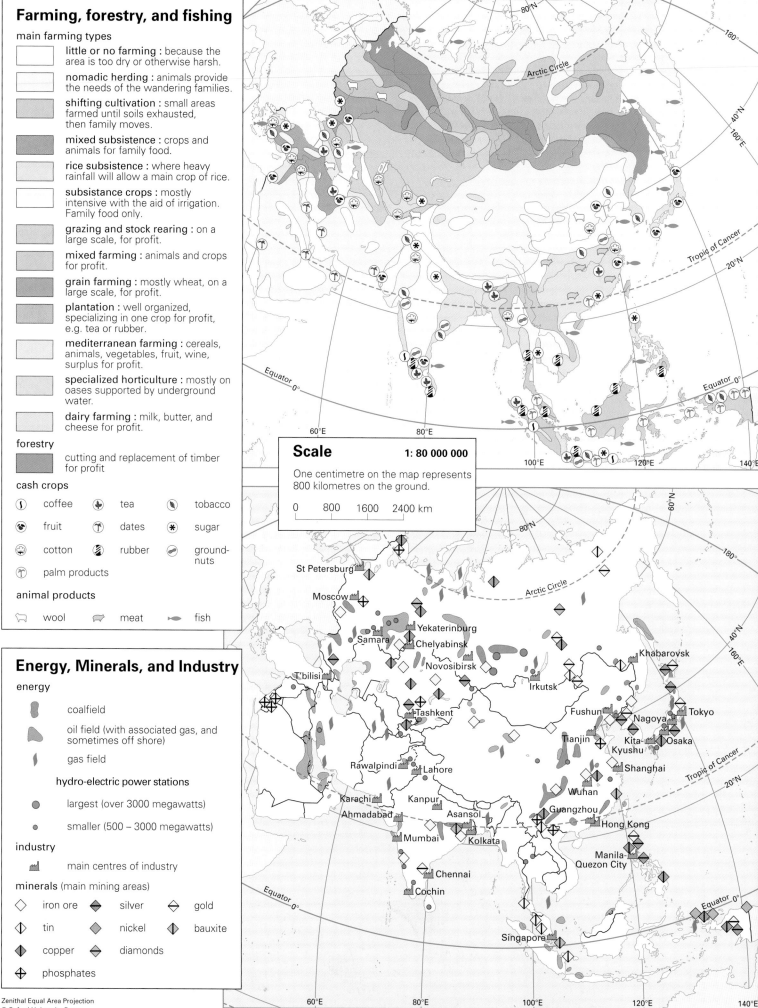

Farming, forestry, and fishing

main farming types

little or no farming : because the area is too dry or otherwise harsh.

nomadic herding : animals provide the needs of the wandering families.

shifting cultivation : small areas farmed until soils exhausted, then family moves.

mixed subsistence : crops and animals for family food.

rice subsistence : where heavy rainfall will allow a main crop of rice.

subsistance crops : mostly intensive with the aid of irrigation. Family food only.

grazing and stock rearing : on a large scale, for profit.

mixed farming : animals and crops for profit.

grain farming : mostly wheat, on a large scale, for profit.

plantation : well organized, specializing in one crop for profit, e.g. tea or rubber.

mediterranean farming : cereals, animals, vegetables, fruit, wine, surplus for profit.

specialized horticulture : mostly on oases supported by underground water.

dairy farming : milk, butter, and cheese for profit.

forestry

cutting and replacement of timber for profit

cash crops

Ⓢ	coffee	✿ tea	tobacco
fruit	🌴 dates	✳ sugar
cotton	rubber	ground-nuts
🌴 palm products		

animal products

wool	meat	fish

Energy, Minerals, and Industry

energy

coalfield

oil field (with associated gas, and sometimes off shore)

gas field

hydro-electric power stations

largest (over 3000 megawatts)

smaller (500 – 3000 megawatts)

industry

main centres of industry

minerals (main mining areas)

◇	iron ore	◈ silver	◇ gold
◁	tin	◆ nickel	bauxite
◆	copper	◆ diamonds	
✢	phosphates		

Scale
1 : 80 000 000

One centimetre on the map represents 800 kilometres on the ground.

0   800   1600   2400 km

St Petersburg
Moscow
Yekaterinburg
Samara
Chelyabinsk
T'bilisi
Novosibirsk
Irkutsk
Khabarovsk
Tashkent
Fushun
Nagoya
Tokyo
Tianjin
Kita-Kyushu
Osaka
Rawalpindi
Lahore
Shanghai
Karachi
Kanpur
Wuhan
Ahmadabad
Asansol
Guangzhou
Mumbai
Kolkata
Hong Kong
Chennai
Manila-Quezon City
Cochin
Singapore

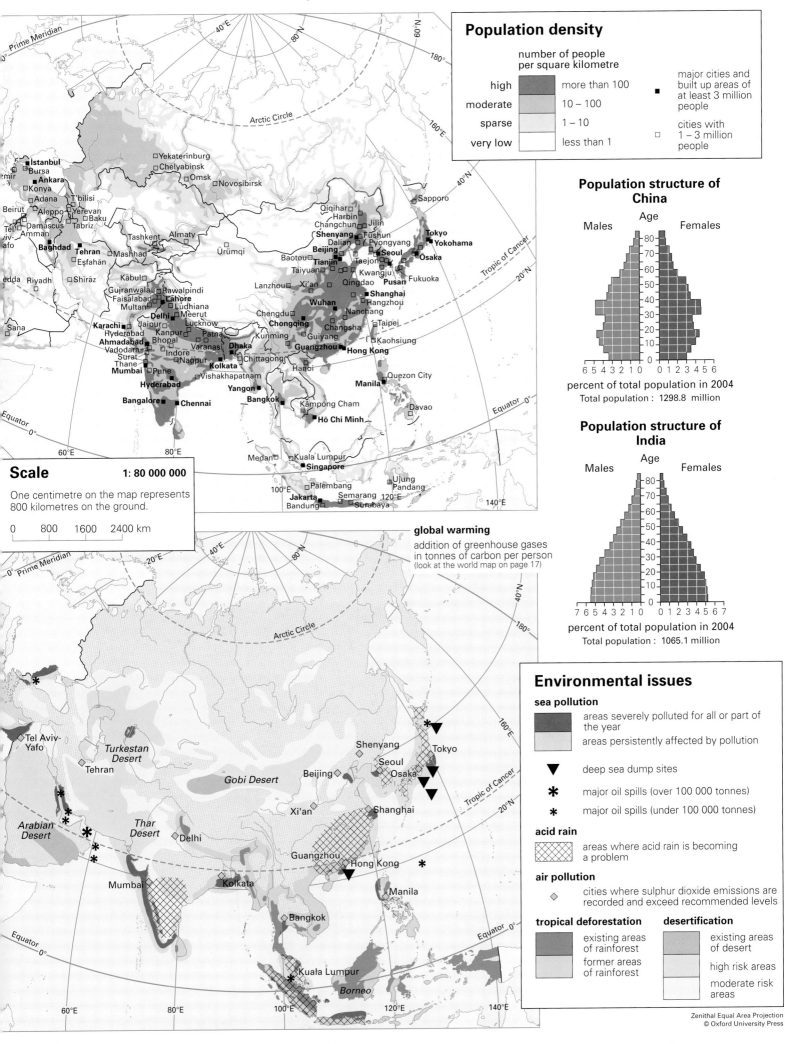

Population density

number of people
per square kilometre

high	more than 100
moderate	10 – 100
sparse	1 – 10
very low	less than 1

■ major cities and built up areas of at least 3 million people

□ cities with 1 – 3 million people

Population structure of China

Age
Males   Females

percent of total population in 2004
Total population : 1298.8 million

Population structure of India

Age
Males   Females

percent of total population in 2004
Total population : 1065.1 million

Scale

1: 80 000 000

One centimetre on the map represents 800 kilometres on the ground.

0   800   1600   2400 km

global warming

addition of greenhouse gases in tonnes of carbon per person
(look at the world map on page 17)

Environmental issues

sea pollution

▓ areas severely polluted for all or part of the year

░ areas persistently affected by pollution

▼ deep sea dump sites

✳ major oil spills (over 100 000 tonnes)

✱ major oil spills (under 100 000 tonnes)

acid rain

▨ areas where acid rain is becoming a problem

air pollution

◇ cities where sulphur dioxide emissions are recorded and exceed recommended levels

tropical deforestation

existing areas of rainforest

former areas of rainforest

desertification

existing areas of desert

high risk areas

moderate risk areas

Zenithal Equal Area Projection
© Oxford University Press

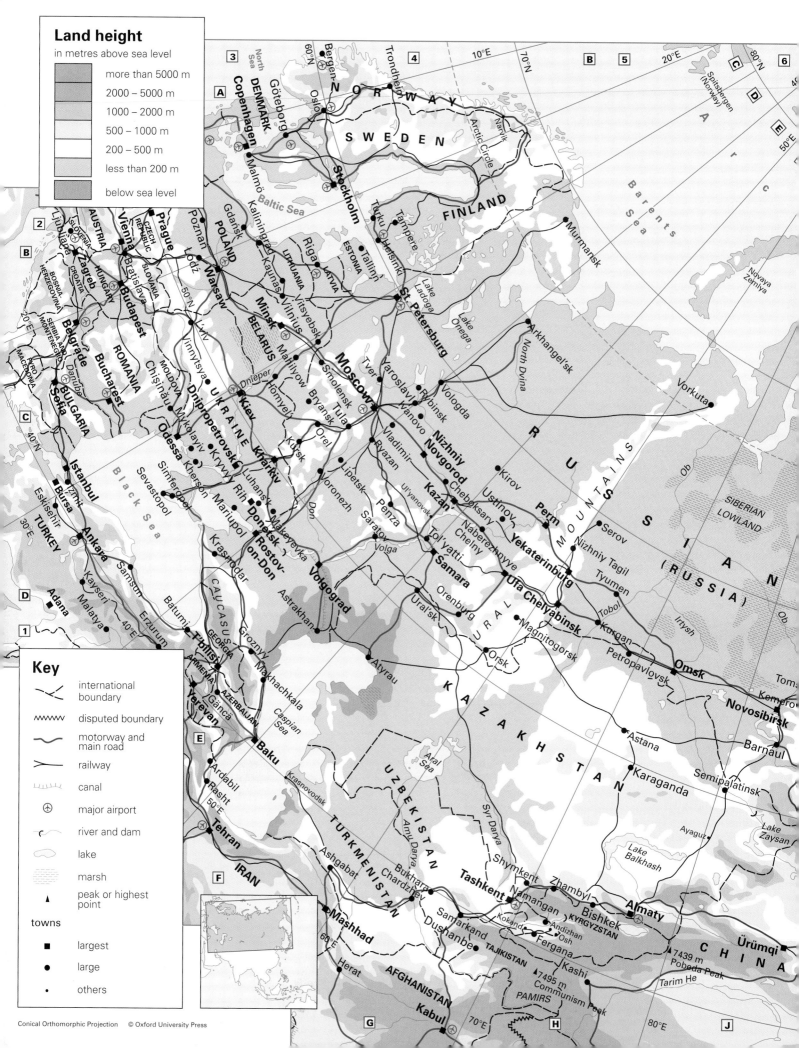

Land height
in metres above sea level

- more than 5000 m
- 2000 – 5000 m
- 1000 – 2000 m
- 500 – 1000 m
- 200 – 500 m
- less than 200 m
- below sea level

Key

- international boundary
- disputed boundary
- motorway and main road
- railway
- canal
- major airport
- river and dam
- lake
- marsh
- peak or highest point

towns

- ■ largest
- ● large
- · others

Conical Orthomorphic Projection   © Oxford University Press

Scale 1: 20 000 000
One centimetre on the map represents
200 kilometres on the ground.

0   200   400   600   800 km

Franz Josef
Land

80°N

180°

70°N

USA

Bering
Strait

170°W

St.
Lawrence

Severnaya
Zemlya

80°E

100°E

120°E

140°E

160°E

180°

60°N

New Siberian
Islands

Wrangel
Island

Bering Sea

170°E

Yenisey

Norilsk

CENTRAL
SIBERIAN
PLATEAU

Verkhoyansk

Lena

Kolyma

Magadan

Kamchatka
Peninsula

160°E

Petropavlovsk-
Kamchatskiy

Yakutsk

Sea of
Okhotsk

50°N

F E D E R A T I O N

Angara

Lena

Tynda

Sakhalin

Kuril
Islands

150°E

Krasnoyarsk

Ust-Kut

Komsomolsk-
on-Amur

Sovetskaya
Gavan

Bratsk

Amur

Blagoveshchensk

Khabarovsk

Lake
Baykal

Shilka

Hegang   Jiamusi

ovokuznetsk

Irkutsk

Chita

Qiqihar

Jixi

Lake
Khanka

Sapporo

Ulan-Ude

Harbin

Vladivostok

Mudanjiang

40°N

Ulan Bator

Kerulen

Changchun   Jilin

Chongjin

Sea of
Japan

Sendai

ALTAI

4226 m

MONGOLIA

Fushun

NORTH KOREA

JAPAN
Tokyo

MOUNTAINS

Shenyang
Anshan

Kyoto   Nagoya

GOBI DESERT

Pyongyang

Osaka

Hami

C H I N A

Seoul
SOUTH
KOREA

130°E

140°E

Hohhot

Beijing

Dalian

100°E

110°E

120°E

Key

---·---	international boundary	
wwwww	disputed boundary	
⌒	motorway and main road	
⟩	railway	
✈	major airport	

⌇	river and dam	
◌	lake	
⠿	marsh	
▲	peak or highest point	

towns

■ largest
● large
· others

Land height

in metres above sea level

more than 5000 m
2000 – 5000 m
1000 – 2000 m
500 – 1000 m
200 – 500 m
less than 200 m
below sea level

Scale 1: 20 000 000

One centimetre on the map represents 200 kilometres on the ground.

0 200 400 600 800 km

Conical Orthomorphic Projection © Oxford University Press

MONGOLIA
GOBI DESERT
ALTAI MOUNTAINS

TURKMENISTAN
UZBEKISTAN
Ashgabat
Mashhad
Bukhara
Chardzhev
Atrek
Zhambyl
Bishkek
Shymkent
Almaty
Namangan
Kokand
Andizhan
Fergana
Osh
KYRGYZSTAN
TIEN SHAN
Pobeda Peak 7439 m
Ürümqi
Turpan Depression −154 m
Hami
Anxi
Jiayuguan
Zhangye
NAN SHAN
Qinghai Hu
Qaidam Pendi (Qaidam Basin)
Xining
Lanzhou
Chengdu

TAJIKISTAN
Samarkand
Dushanbe
Communism Peak 7495 m
PAMIRS
Kashi
Tarim He
TARIM PENDI (Tarim Basin)
CHINA
KUNLUN SHAN
Jinsha Jiang (Yangtze)

IRAN
Zahedan
Herat
HINDU KUSH
Kabul
Khyber Pass
Peshawar
Rawalpindi
Islamabad
JAMMU AND KASHMIR
K2 8611 m
Srinagar
Indus
TIBETAN PLATEAU
Lhasa
Lancang Jiang (Mekong)
Nu Jiang (Salween)
Batang
Gongga Shan 7556 m

AFGHANISTAN
Kandahar
Quetta
Faisalabad
Lahore
Amritsar
Ludhiana
Multan
HIMALAYA
Annapurna 8073 m
Yarlung Zangbo Jiang (Tsangpo)
Mt. Everest 8848 m
Thimphu
Brahmaputra
Dibrugarh
Kunming

PAKISTAN
Sukkur
Bikaner
Meerut
Delhi
Bareilly
NEPAL
Kathmandu
BHUTAN
Guwahati
Imphal

Gwadar
Hyderabad
Karachi
Tropic of Cancer
Gulf of Oman
OMAN
Jaipur
THAR DESERT
Agra
Gwalior
Lucknow
Kanpur
Patna
Allahabad
Varanasi
Asansol
Ganges
BANGLADESH
Dhaka
Chittagong
Mandalay

Kandla
Ahmadabad
Bhopal
Narmada
INDIA
Jabalpur
Jamshedpur
Khulna
Kolkata
MYANMAR (BURMA)
Chiang Mai

Porbandar
Vadodara
Indore
Nagpur
Raipur
Cuttack
Pegu
THAILAND

Arabian Sea
Surat
Mumbai
Pune
Solapur
Godavari
Bay of Bengal
Bassein
Yangon
Moulmein

Kolhapur
Belgaum
Hubli
Krishna
Hyderabad
Vijayawada
Vishakapatnam
Bangkok
Andaman Sea

WESTERN GHATS
DECCAN
Bangalore
Mangalore
Mysore
Chennai
Andaman Islands (India)

Lakshadweep (Laccadive Islands)
Salem
Calicut
Coimbatore
Madurai
Cochin
Jaffna
Trivandrum
Trincomalee
Isthmus of Kra

MALDIVES
SRI LANKA
Colombo
Kandy
Nicobar Islands (India)

Indian Ocean
Banda Aceh
Medan
INDONESIA

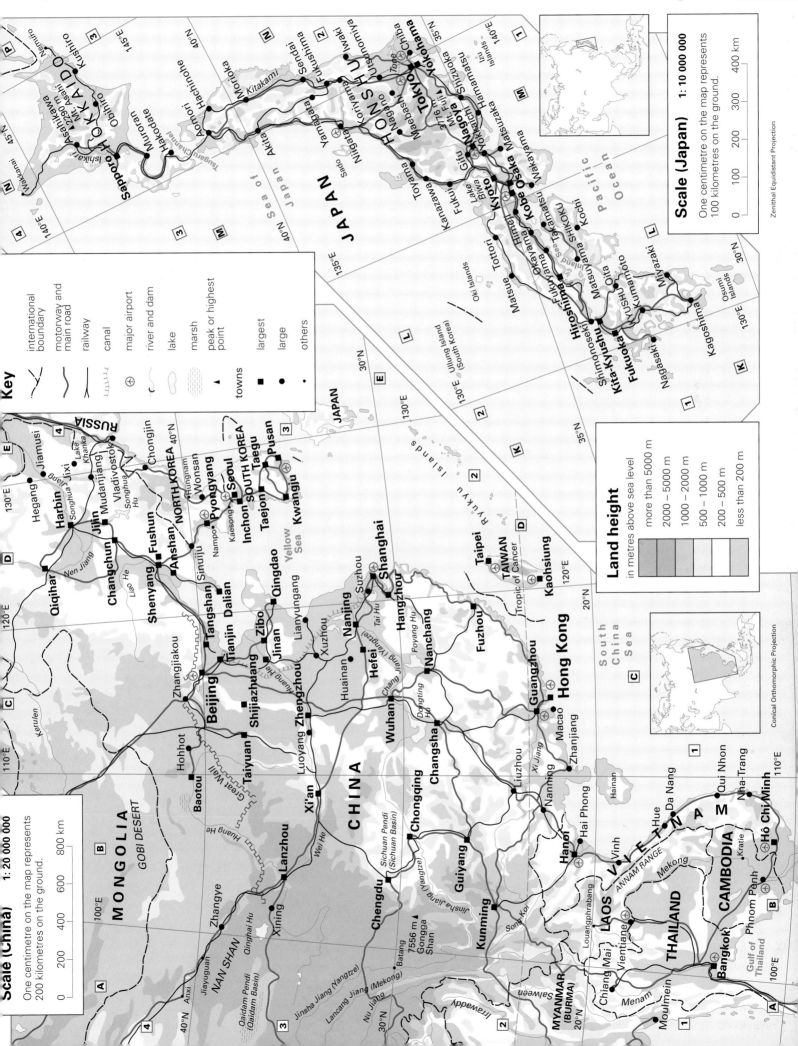

Key

towns
- international boundary
- motorway and main road
- railway
- major airport
- river
- lake
- marsh
- peak or highest point
- largest
- large
- others

Land height
in metres above sea level

more than 2000 m	
1000 – 2000 m	
500 – 1000 m	
200 – 500 m	
less than 200 m	

Scale
1: 20 000 000

One centimetre on the map represents 200 kilometres on the ground.

0   200   400   600   800 km

Conical Orthomorphic Projection   © Oxford University Press

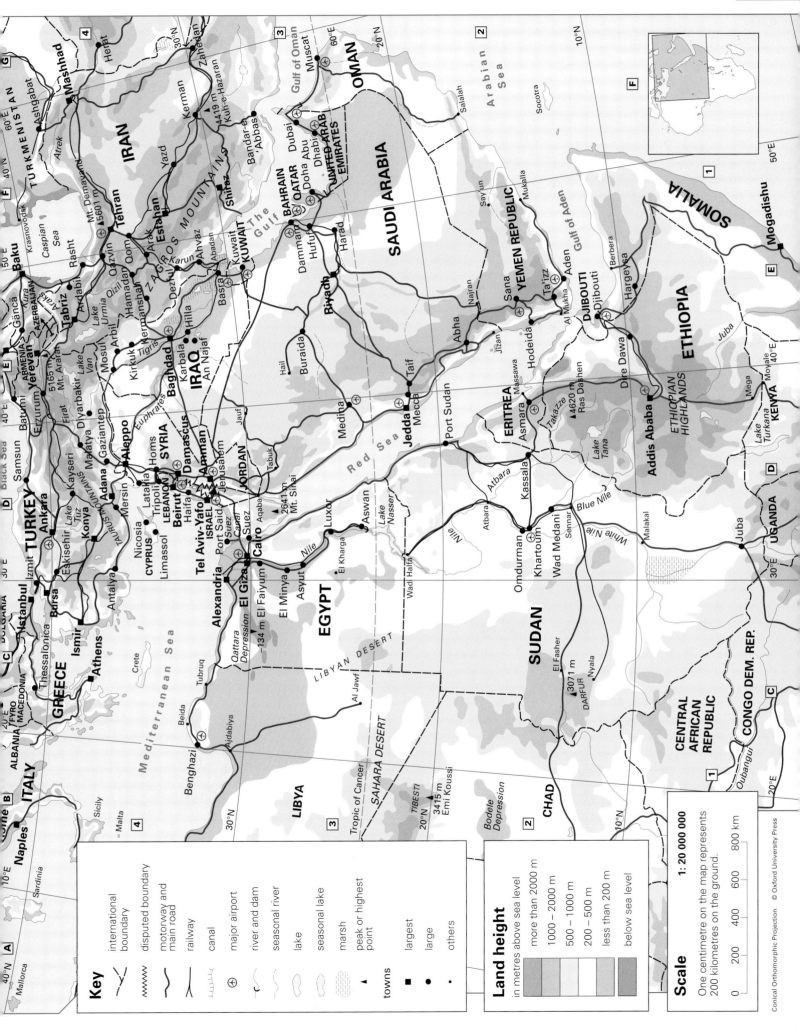

Key

international boundary	
disputed boundary	
motorway and main road	
railway	
canal	
major airport	
river and dam	
seasonal river	
lake	
seasonal lake	
marsh	
peak or highest point	

towns
- largest
- large
- others

Land height

in metres above sea level

- more than 2000 m
- 1000 – 2000 m
- 500 – 1000 m
- 200 – 500 m
- less than 200 m
- below sea level

Scale   1: 20 000 000

One centimetre on the map represents 200 kilometres on the ground.

0   200   400   600   800 km

Conical Orthomorphic Projection   © Oxford University Press

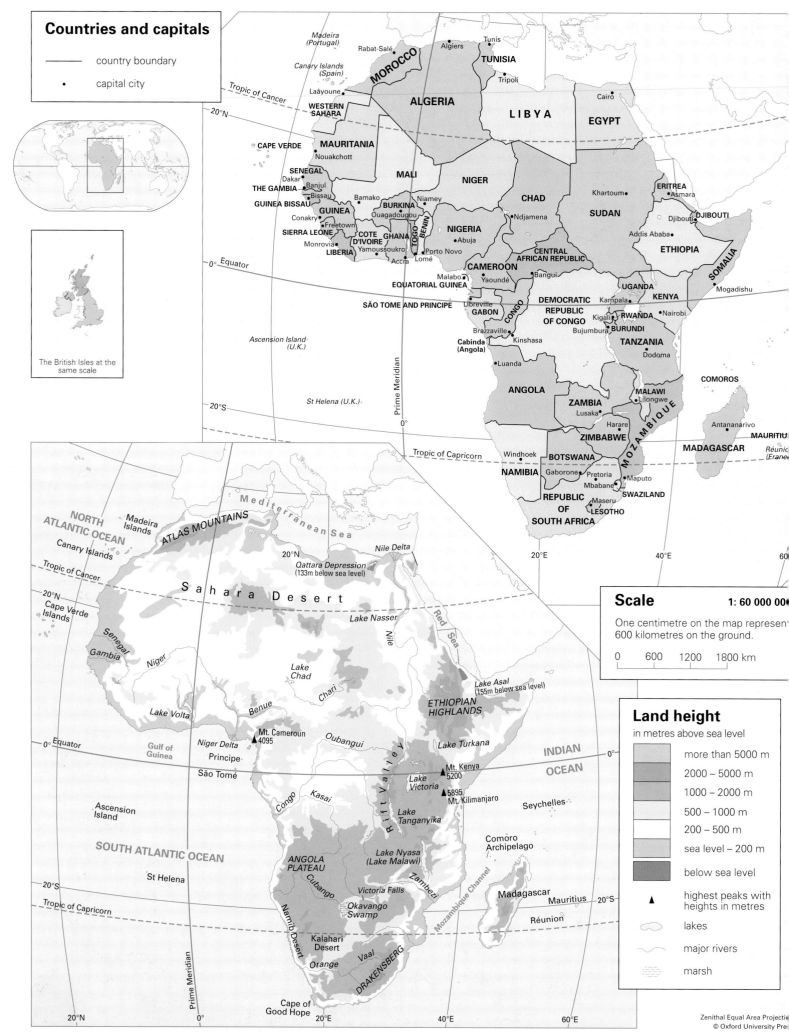

Countries and capitals

——— country boundary

• capital city

The British Isles at the same scale

MOROCCO
Rabat-Salé
Madeira (Portugal)
Canary Islands (Spain)
Laâyoune
WESTERN SAHARA
Algiers
TUNISIA
Tunis
Tripoli
LIBYA
Cairo
EGYPT
ALGERIA

CAPE VERDE
MAURITANIA
Nouakchott
MALI
NIGER
CHAD
Khartoum
SUDAN
ERITREA
Asmara

SENEGAL
Dakar
THE GAMBIA
Banjul
Bissau
GUINEA BISSAU
GUINEA
Conakry
SIERRA LEONE
Freetown
Monrovia
LIBERIA
COTE D'IVOIRE
Bamako
Ouagadougou
BURKINA
Niamey
BENIN
TOGO
GHANA
Yamoussoukro
Accra
Lomé
Porto Novo
NIGERIA
Abuja
Ndjamena
CENTRAL AFRICAN REPUBLIC
Bangui
CAMEROON
Addis Ababa
ETHIOPIA
Djibouti
DJIBOUTI
SOMALIA
Mogadishu

EQUATORIAL GUINEA
Malabo
Yaoundé
SÃO TOME AND PRINCIPE
Libreville
GABON
CONGO
Brazzaville
Kinshasa
Cabinda (Angola)
DEMOCRATIC REPUBLIC OF CONGO
UGANDA
Kampala
Kigali
RWANDA
Bujumbura
BURUNDI
KENYA
Nairobi
TANZANIA
Dodoma

Luanda
ANGOLA
ZAMBIA
Lusaka
MALAWI
Lilongwe
COMOROS
Antananarivo
MADAGASCAR
MAURITIUS
Réunion (France)

Windhoek
NAMIBIA
BOTSWANA
Gaborone
ZIMBABWE
Harare
MOZAMBIQUE
Maputo
SWAZILAND
Mbabane
Pretoria
REPUBLIC OF SOUTH AFRICA
Maseru
LESOTHO

Tropic of Cancer
20°N
Equator
0°
20°S
Tropic of Capricorn
Prime Meridian
0°
20°E
40°E

NORTH ATLANTIC OCEAN
Madeira Islands
ATLAS MOUNTAINS
Mediterranean Sea
Nile Delta
20°N
Qattara Depression (133m below sea level)

Canary Islands
Tropic of Cancer
20°N
Cape Verde Islands

S a h a r a   D e s e r t
Lake Nasser
Nile
Red Sea

Senegal
Gambia
Niger
Lake Chad
Chari
Lake Asal (155m below sea level)
ETHIOPIAN HIGHLANDS

Lake Volta
Benue
Mt. Cameroun 4095
Niger Delta
Principe
Oubangui
Lake Turkana
INDIAN OCEAN

Equator
0°
Gulf of Guinea
São Tomé
Congo
Kasai
Mt. Kenya 5200
Lake Victoria
Mt. Kilimanjaro 5895
0°

Ascension Island
Rift Valley
Lake Tanganyika
Seychelles
Comoro Archipelago

SOUTH ATLANTIC OCEAN
St Helena
ANGOLA PLATEAU
Cubango
Lake Nyasa (Lake Malawi)
Zambezi
Victoria Falls
Madagascar
Mauritius
20°S

Namib Desert
Okavango Swamp
Kalahari Desert
Orange
Vaal
DRAKENSBERG
Réunion
Tropic of Capricorn

Cape of Good Hope
20°N
0°
20°E
40°E
60°E

Scale   1: 60 000 000

One centimetre on the map represen[ts] 600 kilometres on the ground.

0   600   1200   1800 km

Land height
in metres above sea level

more than 5000 m
2000 – 5000 m
1000 – 2000 m
500 – 1000 m
200 – 500 m
sea level – 200 m
below sea level

▲ highest peaks with heights in metres

lakes

major rivers

marsh

Zenithal Equal Area Projection
© Oxford University Pre[ss]

Tamanrasset (1377 m)

[Temperature graph: temperature in degrees Celsius, showing curve rising to peak around July]

[Rainfall graph: rainfall in millimetres, 0–100, months J F M A M J J A S O N D]

Douala (8 m)

[Temperature graph: temperature in degrees Celsius]

[Rainfall graph: rainfall in millimetres, scale 0–750, months J F M A M J J A S O N D]

Durban (5 m)

[Temperature graph: temperature in degrees Celsius]

[Rainfall graph: rainfall in millimetres, 0–150, months J F M A M J J A S O N D]

Tamanrasset
Mean annual rainfall: 54 mm
Mean January temperature: 12.5°C
Mean July temperature: 28.5°C

Douala
Mean annual rainfall: 4027 mm
Mean January temperature: 26.5°C
Mean July temperature: 24.5°C

Durban
Mean annual rainfall: 1008 mm
Mean January temperature: 24.0°C
Mean July temperature: 16.5°C

Climatic regions

Hot tropical rainy
- rain all year
- monsoon
- dry in winter

Very dry
- with no reliable rain
- with a little rain

Influenced by the sea: warm summers, mild winters
- with dry summers (Mediterranean type)
- with dry winters
- with no dry season

Mountain
- height of the land strongly affects the climate

Ocean currents
- → warm
- → cold

Climate recording stations
- • climate recording stations for which graphs are shown

Scale   1 : 60 000 000

One centimetre on the map represents 600 kilometres on the ground.

0   600   1200   1800 km

Ecosystems

Vegetation types are those which would occur naturally without interference by people.

- tropical rain forest
- tropical grasslands (savannah)
- evergreen trees and shrubs
- thorn forest
- temperate grasslands
- semi-desert
- desert
- mountains

More information about these ecosystems can be found on page 8.

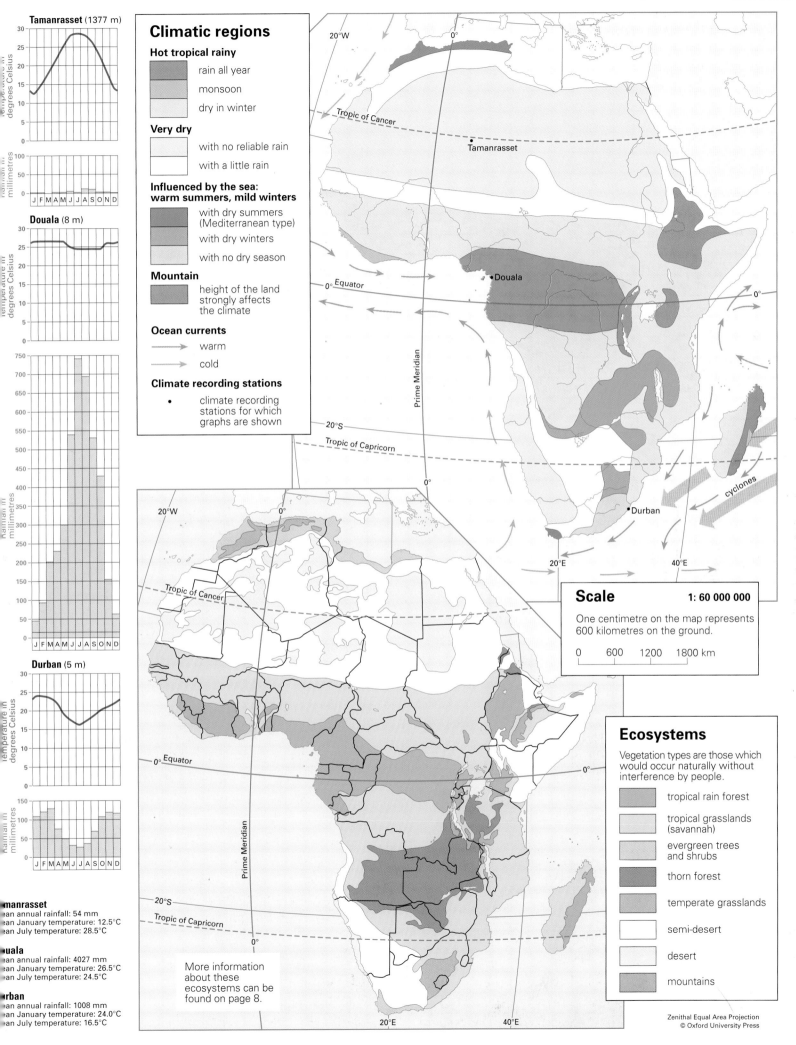

Zenithal Equal Area Projection
© Oxford University Press

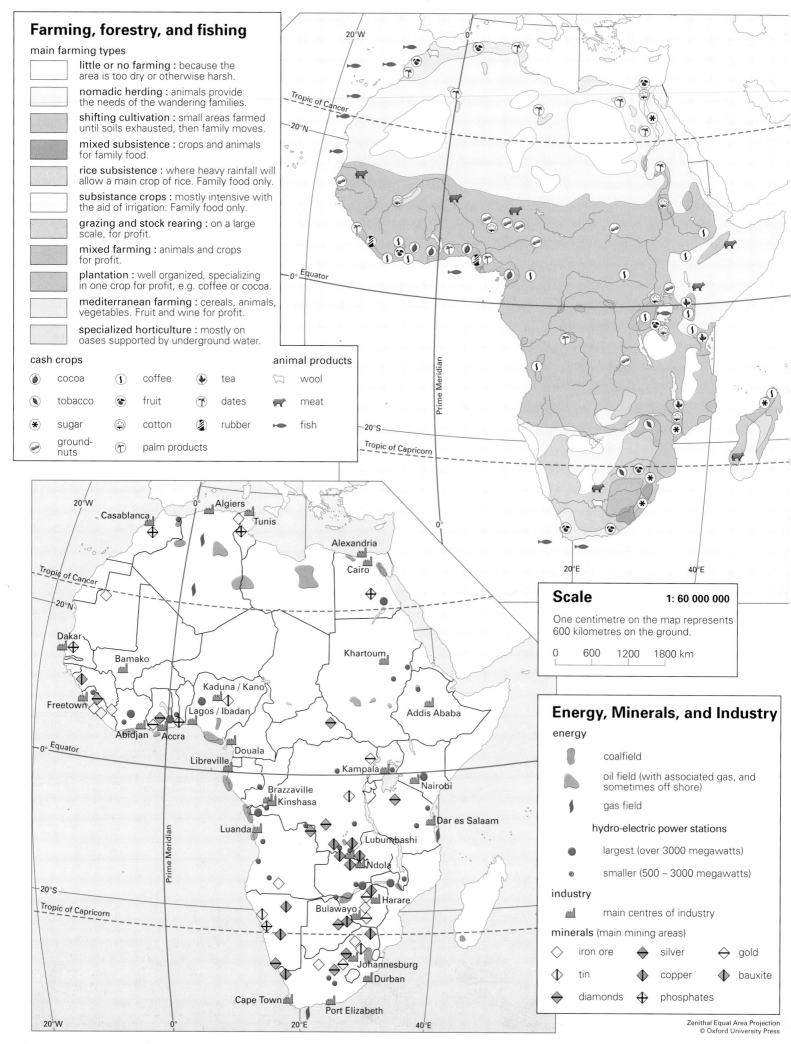

Farming, forestry, and fishing

main farming types

little or no farming : because the area is too dry or otherwise harsh.

nomadic herding : animals provide the needs of the wandering families.

shifting cultivation : small areas farmed until soils exhausted, then family moves.

mixed subsistence : crops and animals for family food.

rice subsistence : where heavy rainfall will allow a main crop of rice. Family food only.

subsistance crops : mostly intensive with the aid of irrigation. Family food only.

grazing and stock rearing : on a large scale, for profit.

mixed farming : animals and crops for profit.

plantation : well organized, specializing in one crop for profit, e.g. coffee or cocoa.

mediterranean farming : cereals, animals, vegetables. Fruit and wine for profit.

specialized horticulture : mostly on oases supported by underground water.

cash crops

- cocoa
- tobacco
- sugar
- ground-nuts
- coffee
- fruit
- cotton
- palm products
- tea
- dates
- rubber

animal products

- wool
- meat
- fish

Scale 1: 60 000 000

One centimetre on the map represents 600 kilometres on the ground.

0   600   1200   1800 km

Energy, Minerals, and Industry

energy

- coalfield
- oil field (with associated gas, and sometimes off shore)
- gas field

hydro-electric power stations

- largest (over 3000 megawatts)
- smaller (500 – 3000 megawatts)

industry

- main centres of industry

minerals (main mining areas)

- iron ore
- tin
- diamonds
- silver
- copper
- phosphates
- gold
- bauxite

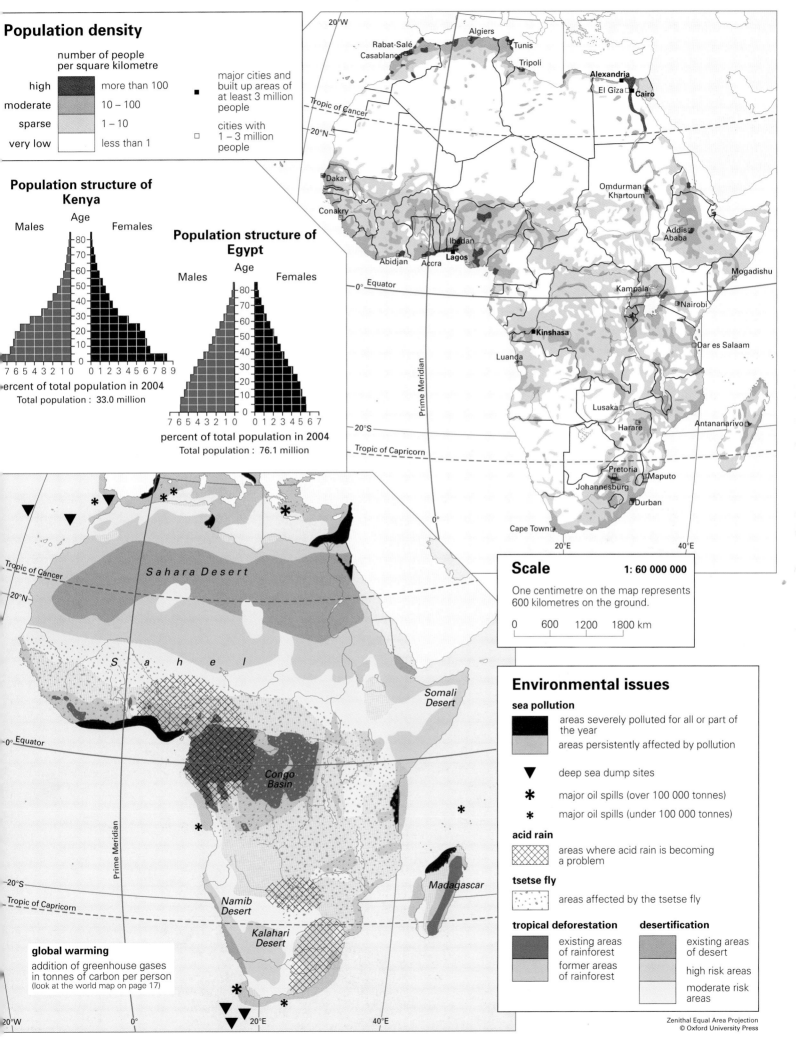

Population density

number of people per square kilometre

high	more than 100
moderate	10 – 100
sparse	1 – 10
very low	less than 1

■ major cities and built up areas of at least 3 million people

□ cities with 1 – 3 million people

Population structure of Kenya

Age

Males   Females

80 70 60 50 40 30 20 10 0

7 6 5 4 3 2 1 0   0 1 2 3 4 5 6 7 8 9

percent of total population in 2004
Total population : 33.0 million

Population structure of Egypt

Age

Males   Females

80 70 60 50 40 30 20 10 0

7 6 5 4 3 2 1 0   0 1 2 3 4 5 6 7

percent of total population in 2004
Total population : 76.1 million

Scale   1: 60 000 000

One centimetre on the map represents 600 kilometres on the ground.

0   600   1200   1800 km

global warming

addition of greenhouse gases in tonnes of carbon per person
(look at the world map on page 17)

Environmental issues

sea pollution

▬ areas severely polluted for all or part of the year

areas persistently affected by pollution

▼ deep sea dump sites

✳ major oil spills (over 100 000 tonnes)

✳ major oil spills (under 100 000 tonnes)

acid rain

▨ areas where acid rain is becoming a problem

tsetse fly

areas affected by the tsetse fly

tropical deforestation

existing areas of rainforest

former areas of rainforest

desertification

existing areas of desert

high risk areas

moderate risk areas

Zenithal Equal Area Projection
© Oxford University Press

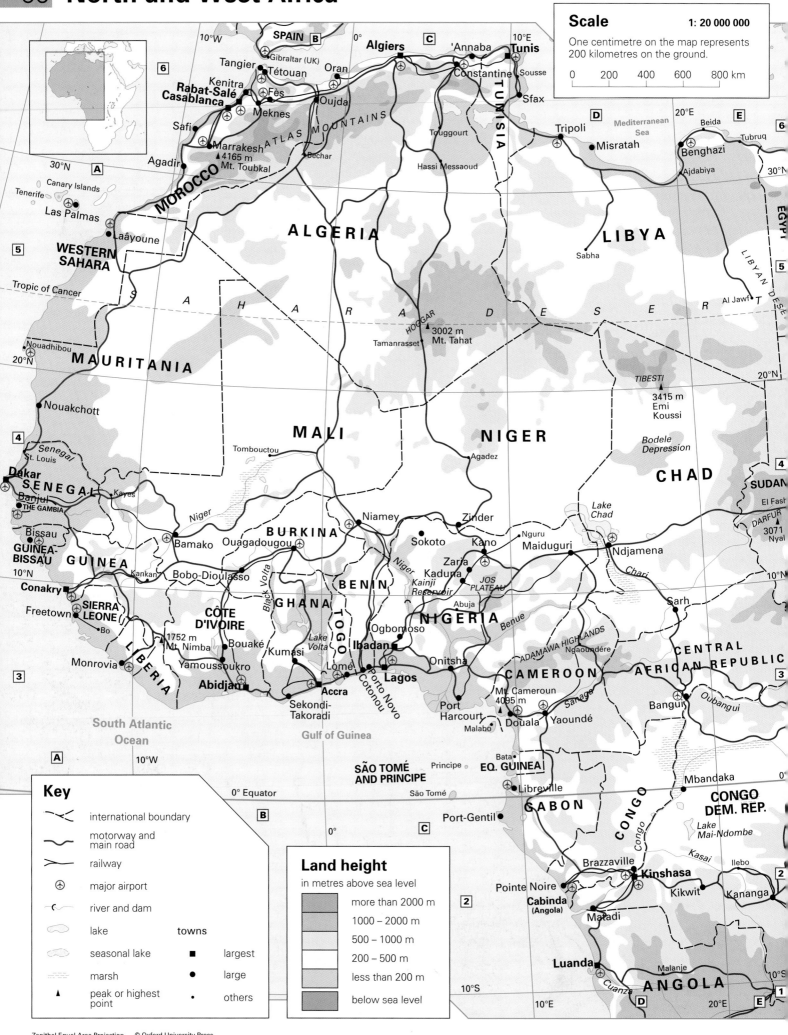

Scale   1: 20 000 000

One centimetre on the map represents 200 kilometres on the ground.

0   200   400   600   800 km

Key

	international boundary
	motorway and main road
	railway
⊕	major airport
	river and dam
	lake
	seasonal lake
	marsh
▲	peak or highest point

towns

■	largest
●	large
·	others

Land height

in metres above sea level

	more than 2000 m
	1000 – 2000 m
	500 – 1000 m
	200 – 500 m
	less than 200 m
	below sea level

Zenithal Equal Area Projection   © Oxford University Press

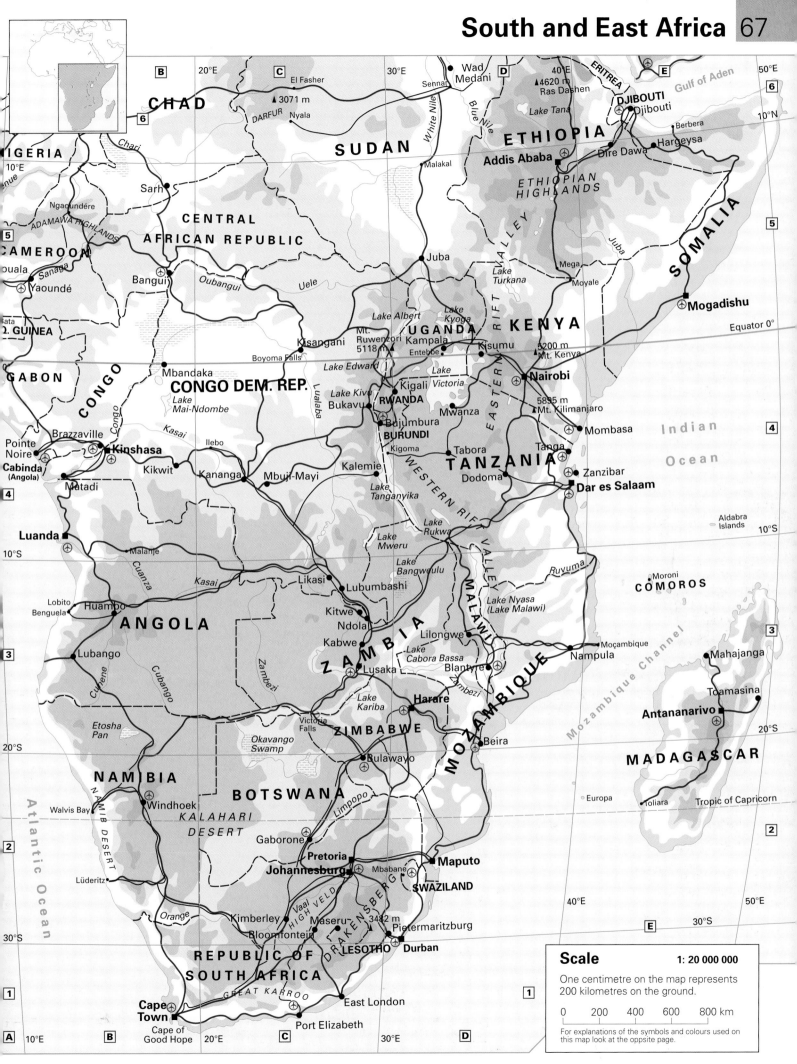

B 20°E C 30°E D 40°E ERITREA E 50°E

6 **CHAD** El Fasher Wad Medani ▲4620 m Ras Dashen Gulf of Aden 6

SUDAN ▲3071 m DARFUR Nyala Sennar Lake Tana **DJIBOUTI** Djibouti 10°N

Chari Malakal White Nile Blue Nile **ETHIOPIA** Berbera

NIGERIA ADAMAWA HIGHLANDS **CENTRAL** **Addis Ababa** Dire Dawa Hargeysa

10°E Ngaoundéré **AFRICAN REPUBLIC** ETHIOPIAN HIGHLANDS

5 CAMEROON Sarh 5

ouala Sanaga Yaoundé Bangui Oubangui Uele Juba Lake Turkana Mega Moyale **SOMALIA**

Q. GUINEA Boyoma Falls Kisangani Lake Albert Lake Kyoga **UGANDA** **KENYA** Equator 0°

4 GABON **CONGO** Mbandaka **CONGO DEM. REP.** Mt. Ruwenzori 5118 m Kampala Entebbe Kisumu 5200 m Mt. Kenya **Nairobi** 5895 m Mt. Kilimanjaro Indian **Mogadishu** 4

Congo Kasai Lake Mai-Ndombe Lake Edward Kigali **RWANDA** Bukavu Lake Kivu Lake Victoria Mombasa Ocean

Brazzaville Pointe Noire **Kinshasa** Ilebo Kikwit Bujumbura **BURUNDI** Kigoma Mwanza Tabora **TANZANIA** Tanga Zanzibar

4 **Cabinda** (Angola) Matadi Kananga Mbuji-Mayi Kalemie Lake Tanganyika WESTERN RIFT Dodoma **Dar es Salaam** Aldabra Islands 10°S

Luanda Malanje Likasi Lubumbashi Lake Mweru Lake Rukwa VALLEY Ruvuma Moroni

10°S Cuanza Kasai Kitwe Lake Bangweulu **COMOROS**

3 Lobito Benguela Huambo **ANGOLA** Zambezi Ndola Kabwe **ZAMBIA** Lake Cabora Bassa Lilongwe **MALAWI** Lake Nyasa (Lake Malawi) Nampula Moçambique Mahajanga 3

Lubango Cubango Lusaka Blantyre Zambezi **MOZAMBIQUE** Mozambique Channel Toamasina

Cunene Cubango Zambezi Lake Kariba **Harare** **Antananarivo** 20°S

20°S Etosha Pan Okavango Swamp Victoria Falls **ZIMBABWE** Beira **MADAGASCAR**

2 **NAMIBIA** Bulawayo 2

Walvis Bay NAMIB DESERT **BOTSWANA** Limpopo Europa Tropic of Capricorn

Windhoek KALAHARI DESERT Gaborone Limpopo Toliara

Lüderitz **Pretoria** **Maputo** Mbabane **SWAZILAND** 40°E 50°E

Orange **Johannesburg** HIGH VELD Vaal DRAKENSBERG 3482 m 30°S

1 Kimberley Maseru **Durban** 1

Bloemfontein Pietermaritzburg **LESOTHO**

REPUBLIC OF SOUTH AFRICA GREAT KARROO

Cape Town East London

Cape of Good Hope Port Elizabeth

A 10°E B 20°E C 30°E D

Atlantic Ocean

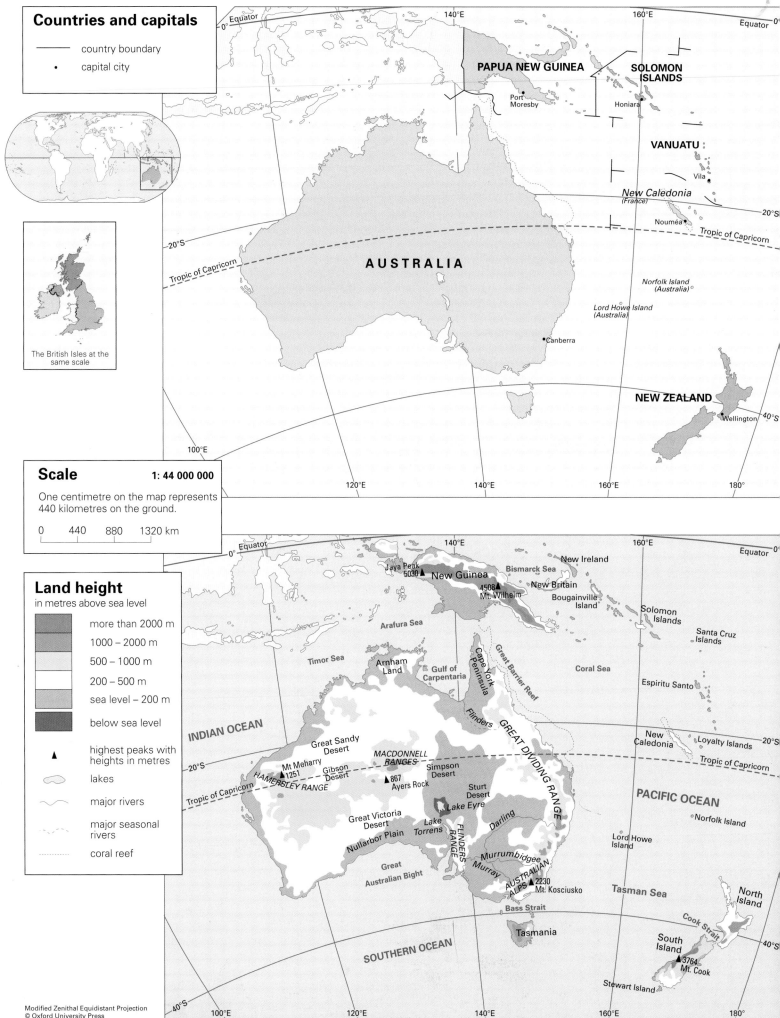

Countries and capitals

— country boundary
• capital city

The British Isles at the same scale

Scale   1: 44 000 000

One centimetre on the map represents 440 kilometres on the ground.

0   440   880   1320 km

Land height
in metres above sea level

more than 2000 m
1000 – 2000 m
500 – 1000 m
200 – 500 m
sea level – 200 m
below sea level

▲   highest peaks with heights in metres
lakes
major rivers
major seasonal rivers
coral reef

PAPUA NEW GUINEA

Port Moresby

SOLOMON ISLANDS

Honiara

VANUATU

Vila

New Caledonia (France)

Nouméa

Norfolk Island (Australia)

Lord Howe Island (Australia)

AUSTRALIA

Canberra

NEW ZEALAND

Wellington

Equator
140°E
160°E
Equator
20°S
Tropic of Capricorn
Tropic of Capricorn
20°S
40°S
100°E
120°E
140°E
160°E
180°

Jaya Peak 5030▲
New Guinea
Bismarck Sea
New Ireland
4508▲ Mt. Wilhelm
New Britain
Bougainville Island
Solomon Islands
Santa Cruz Islands
Arafura Sea
Timor Sea
Arnham Land
Gulf of Carpentaria
Cape York Peninsula
Great Barrier Reef
Coral Sea
Espiritu Santo
INDIAN OCEAN
Flinders
GREAT DIVIDING RANGE
New Caledonia
Loyalty Islands
Great Sandy Desert
MACDONNELL RANGES
Simpson Desert
Mt Meharry ▲1251
Gibson Desert
HAMERSLEY RANGE
▲867 Ayers Rock
Sturt Desert
Tropic of Capricorn
20°S
PACIFIC OCEAN
Lake Eyre
Great Victoria Desert
Lake Torrens
Darling
Norfolk Island
Nullarbor Plain
FLINDERS RANGE
Murrumbidgee
Lord Howe Island
Great Australian Bight
Murray
AUSTRALIAN ALPS ▲2230 Mt. Kosciusko
Tasman Sea
North Island
Bass Strait
Cook Strait
SOUTHERN OCEAN
Tasmania
South Island
3764▲ Mt. Cook
Stewart Island

Modified Zenithal Equidistant Projection
© Oxford University Press

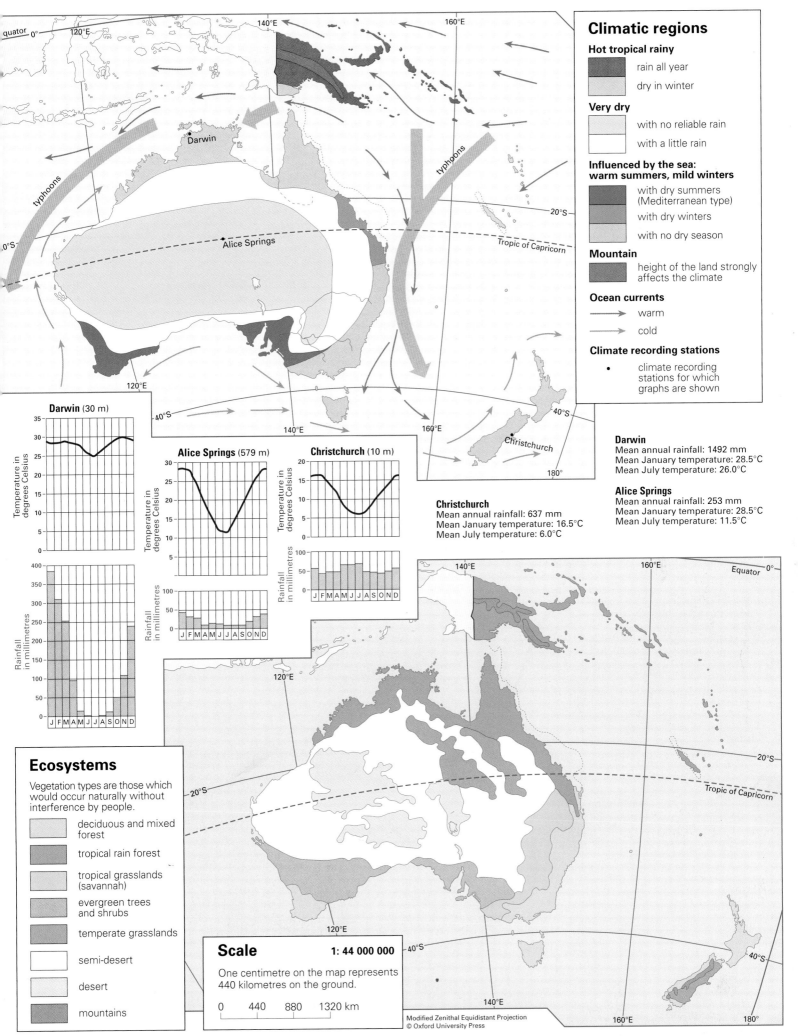

Climatic regions

Hot tropical rainy
- rain all year
- dry in winter

Very dry
- with no reliable rain
- with a little rain

Influenced by the sea: warm summers, mild winters
- with dry summers (Mediterranean type)
- with dry winters
- with no dry season

Mountain
- height of the land strongly affects the climate

Ocean currents
- → warm
- → cold

Climate recording stations
- • climate recording stations for which graphs are shown

Darwin (30 m)

Temperature in degrees Celsius

Rainfall in millimetres

Alice Springs (579 m)

Temperature in degrees Celsius

Rainfall in millimetres

Christchurch (10 m)

Temperature in degrees Celsius

Rainfall in millimetres

Darwin
Mean annual rainfall: 1492 mm
Mean January temperature: 28.5°C
Mean July temperature: 26.0°C

Alice Springs
Mean annual rainfall: 253 mm
Mean January temperature: 28.5°C
Mean July temperature: 11.5°C

Christchurch
Mean annual rainfall: 637 mm
Mean January temperature: 16.5°C
Mean July temperature: 6.0°C

Ecosystems

Vegetation types are those which would occur naturally without interference by people.

- deciduous and mixed forest
- tropical rain forest
- tropical grasslands (savannah)
- evergreen trees and shrubs
- temperate grasslands
- semi-desert
- desert
- mountains

Scale   1: 44 000 000

One centimetre on the map represents 440 kilometres on the ground.

0   440   880   1320 km

Modified Zenithal Equidistant Projection
© Oxford University Press

Farming, forestry, and fishing

main farming types

	little or no farming : because the area is too dry or otherwise harsh.
	shifting cultivation : small areas farmed until soils exhausted, then family moves.
	mixed subsistence : crops and animals for family food.
	grazing and stock rearing : on a large scale, for profit.
	intensive grazing : fattening of lambs, mainly for meat, and of beef cattle. All for profit.
	mixed farming : animals and crops for profit.
	grain farming : mostly wheat but also other cereals, for profit.
	plantation : well organized, specializing in one crop for profit, e.g. sugar or cocoa.
	specialized horticulture : mostly supported by irrigation.
	dairy farming : milk, butter, and cheese for profit. Also lamb fattening in New Zealand.

forestry

forestry for profit

cash crops

- cocoa
- coffee
- fruit
- sugar
- cotton
- rice
- palm products

animal products

- wool
- meat
- fish

area irrigated by the River Murray Scheme

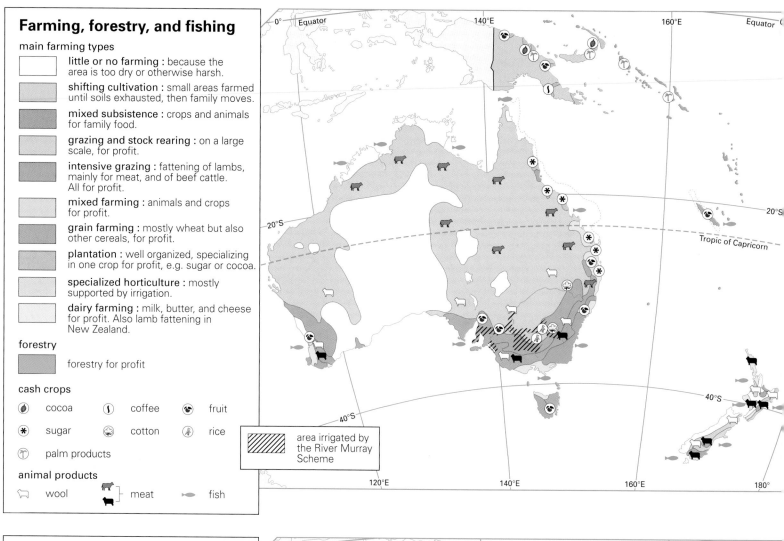

Energy, Minerals, and Industry

energy

- coalfield
- oil field (with associated gas, and sometimes off shore)
- gas field

hydro-electric power stations

- largest (over 3000 megawatts)
- smaller (500 – 3000 megawatts)

industry

- main centres of industry

minerals (main mining areas)

- silver
- gold
- tin
- copper
- bauxite
- nickel
- zinc
- lead
- uranium
- diamonds
- iron ore (iron sands in New Zealand)

Australian underground water supplies

	areas where artesian water is generally available
	areas where artesian water is available in places

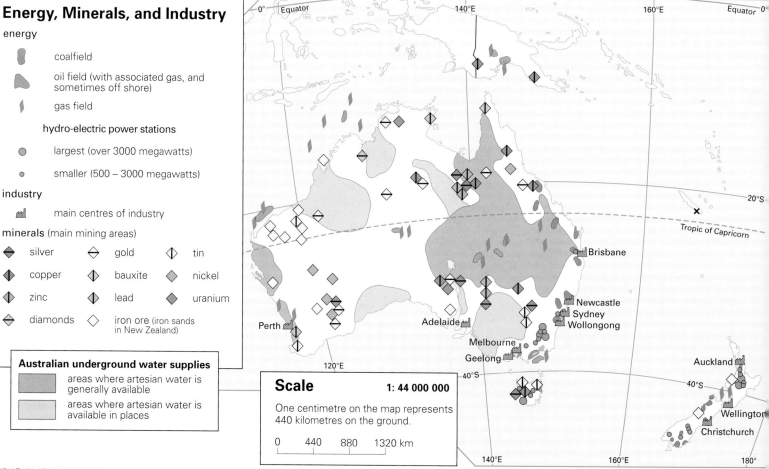

Scale 1: 44 000 000

One centimetre on the map represents 440 kilometres on the ground.

0   440   880   1320 km

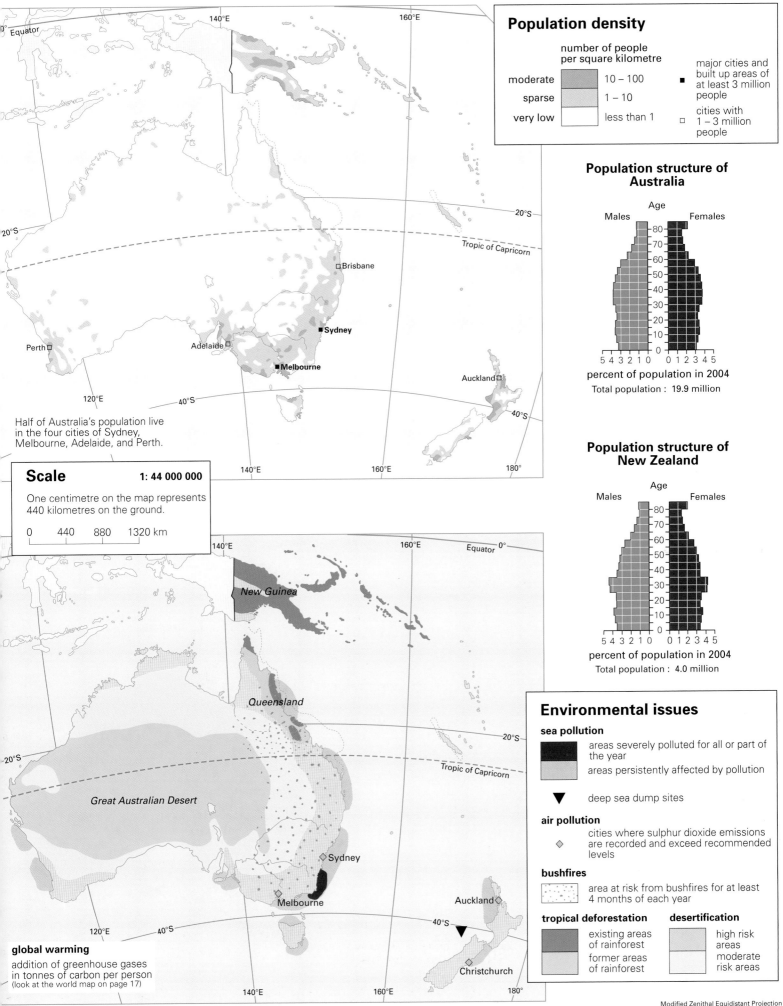

Population density

number of people per square kilometre

moderate	10 – 100
sparse	1 – 10
very low	less than 1

■ major cities and built up areas of at least 3 million people

□ cities with 1 – 3 million people

Tropic of Capricorn

Brisbane

Sydney

Perth

Adelaide

Melbourne

Auckland

Half of Australia's population live in the four cities of Sydney, Melbourne, Adelaide, and Perth.

Population structure of Australia

Age
Males   Females

percent of population in 2004
Total population : 19.9 million

Population structure of New Zealand

Age
Males   Females

percent of population in 2004
Total population : 4.0 million

Scale   1: 44 000 000

One centimetre on the map represents 440 kilometres on the ground.

0   440   880   1320 km

New Guinea

Queensland

Great Australian Desert

Sydney

Melbourne

Auckland

Christchurch

Environmental issues

sea pollution

■ areas severely polluted for all or part of the year

areas persistently affected by pollution

▼ deep sea dump sites

air pollution

◇ cities where sulphur dioxide emissions are recorded and exceed recommended levels

bushfires

area at risk from bushfires for at least 4 months of each year

tropical deforestation

existing areas of rainforest

former areas of rainforest

desertification

high risk areas

moderate risk areas

global warming

addition of greenhouse gases in tonnes of carbon per person
(look at the world map on page 17)

Modified Zenithal Equidistant Projection
© Oxford University Press

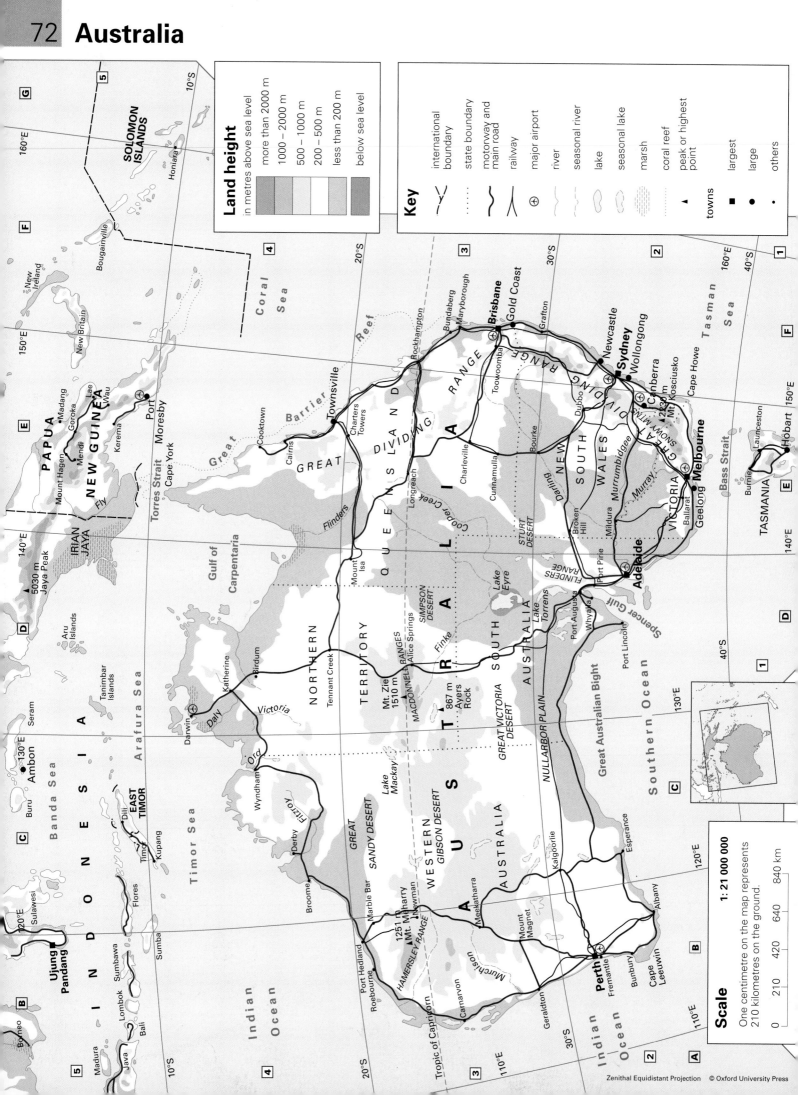

Land height
in metres above sea level

more than 2000 m
1000 – 2000 m
500 – 1000 m
200 – 500 m
less than 200 m
below sea level

Key

international boundary	
state boundary	
motorway and main road	
railway	
major airport	⊕
river	
seasonal river	
lake	
seasonal lake	
marsh	
coral reef	
peak or highest point	▲
towns	■ largest ● large ・ others

Scale   1: 21 000 000

One centimetre on the map represents
210 kilometres on the ground.

0   210   420   640   840 km

Zenithal Equidistant Projection   © Oxford University Press

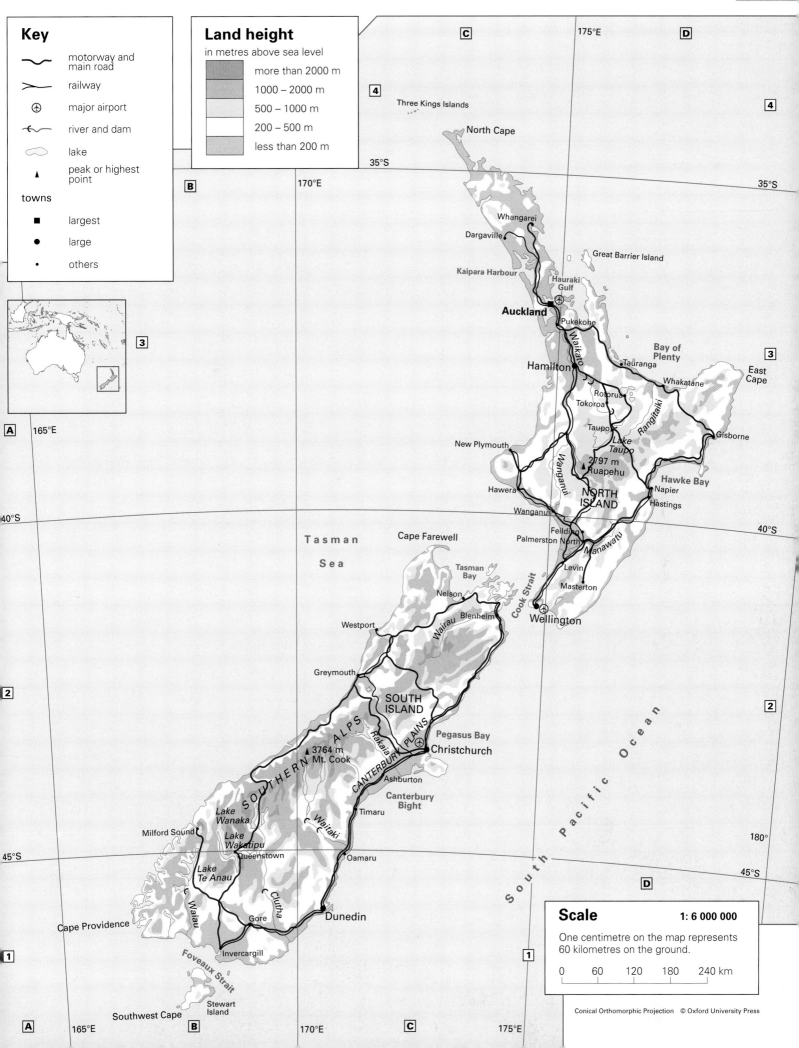

Key

∿	motorway and main road
⌒	railway
⊕	major airport
⌒	river and dam
◯	lake
▲	peak or highest point

towns

■	largest
●	large
•	others

Land height

in metres above sea level

	more than 2000 m
	1000 – 2000 m
	500 – 1000 m
	200 – 500 m
	less than 200 m

Scale 1: 6 000 000

One centimetre on the map represents 60 kilometres on the ground.

0   60   120   180   240 km

Conical Orthomorphic Projection   © Oxford University Press

Three Kings Islands
North Cape
Whangarei
Dargaville
Kaipara Harbour
Great Barrier Island
Hauraki Gulf
Auckland
Pukekohe
Waikato
Bay of Plenty
Hamilton
Tauranga
Rotorua
Whakatane
East Cape
Tokoroa
Rangitaiki
Taupo
Gisborne
New Plymouth
Lake Taupo
Wanganui
2797 m ▲ Ruapehu
Hawke Bay
Hawera
Napier
NORTH ISLAND
Hastings
Wanganui
Feilding
Palmerston North
Manawatu
Levin
Cook Strait
Masterton
Tasman Sea
Cape Farewell
Wellington
Tasman Bay
Nelson
Westport
Wairau
Blenheim
Greymouth
SOUTH ISLAND
Pegasus Bay
SOUTHERN ALPS
Rakaia
CANTERBURY PLAINS
Christchurch
3764 m ▲ Mt. Cook
Ashburton
Canterbury Bight
Lake Wanaka
Timaru
Waitaki
Milford Sound
Lake Wakatipu
Queenstown
Oamaru
South Pacific Ocean
Lake Te Anau
Clutha
Cape Providence
Waiau
Gore
Dunedin
Invercargill
Foveaux Strait
Southwest Cape
Stewart Island

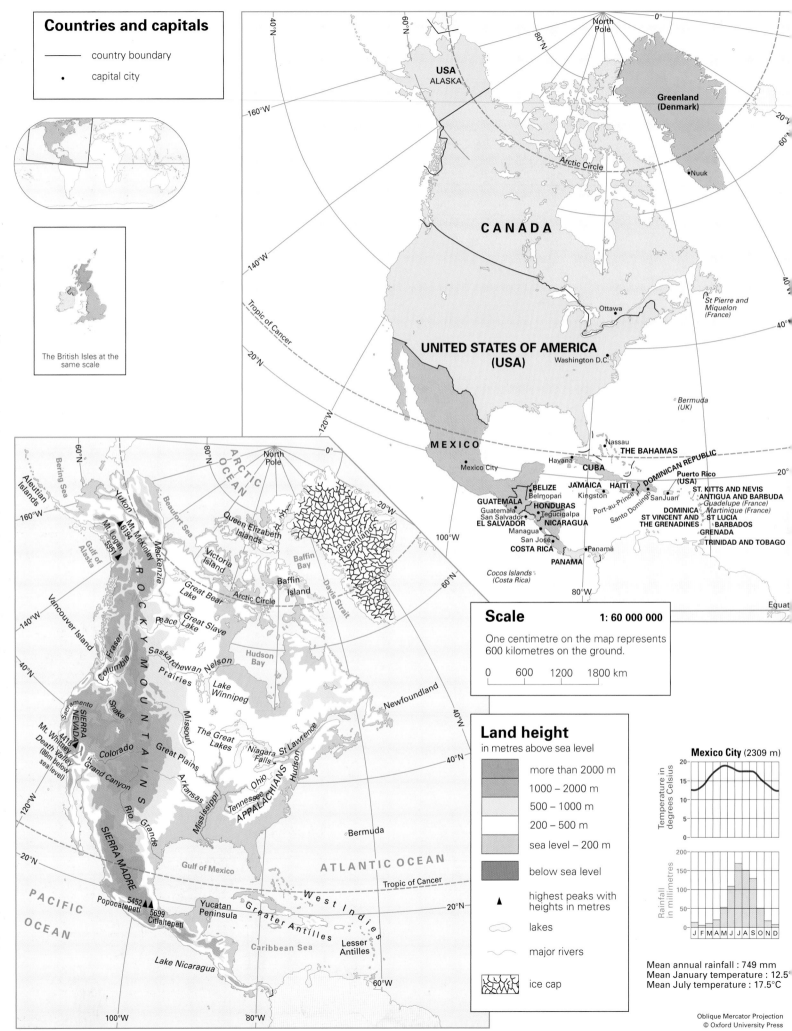

Countries and capitals

— country boundary

• capital city

The British Isles at the same scale

USA
ALASKA

North Pole

Greenland
(Denmark)

Arctic Circle

Nuuk

CANADA

Ottawa

St Pierre and
Miquelon
(France)

UNITED STATES OF AMERICA
(USA)

Washington D.C.

Bermuda
(UK)

MEXICO

Nassau   THE BAHAMAS

Mexico City   Havana   CUBA   DOMINICAN REPUBLIC

BELIZE   JAMAICA   HAITI   Puerto Rico
(USA)
San Juan   ST. KITTS AND NEVIS
Belmopan   Kingston   Port-au-Prince   ANTIGUA AND BARBUDA
GUATEMALA   HONDURAS   Santo Domingo   Guadelupe (France)
Guatemala   Tegucigalpa   DOMINICA   Martinique (France)
San Salvador   ST VINCENT AND   ST LUCIA
EL SALVADOR   NICARAGUA   THE GRENADINES   BARBADOS
Managua   GRENADA
San José   TRINIDAD AND TOBAGO
COSTA RICA   Panamá

PANAMA

Cocos Islands
(Costa Rica)

Equat

Scale   1: 60 000 000

One centimetre on the map represents
600 kilometres on the ground.

0   600   1200   1800 km

Land height
in metres above sea level

more than 2000 m

1000 – 2000 m

500 – 1000 m

200 – 500 m

sea level – 200 m

below sea level

▲ highest peaks with
heights in metres

lakes

major rivers

ice cap

Aleutian
Islands

Bering Sea

Gulf of
Alaska

ARCTIC OCEAN

North Pole

Queen Elizabeth
Islands

Greenland

Baffin
Bay

Beaufort Sea

Yukon   Mt McKinley
6194
Mt Logan
5951

Victoria
Island

Vancouver Island

Mackenzie

ROCKY MOUNTAINS

Great Bear
Lake

Great Slave
Lake

Baffin
Island

Arctic Circle

Davis Strait

Fraser

Columbia

Peace

Saskatchewan   Nelson

Prairies

Hudson
Bay

Newfoundland

Snake

Sacramento

SIERRA
NEVADA
Mt Whitney
4418
Death Valley
(86m below
sea level)

Colorado

Grand Canyon

Missouri

Lake
Winnipeg

The Great
Lakes

Niagara   St Lawrence
Falls

Great Plains

Ohio   APPALACHIANS   Hudson

Arkansas   Tennessee

Rio Grande

Mississippi

SIERRA MADRE

Gulf of Mexico

Bermuda

ATLANTIC OCEAN

Tropic of Cancer

PACIFIC OCEAN

Popocatepetl
5452

Citlaltepetl
5699

Yucatan
Peninsula

West Indies

Greater Antilles

Caribbean Sea

Lesser
Antilles

Lake Nicaragua

Mexico City (2309 m)

Temperature in degrees Celsius

Rainfall in millimetres

J F M A M J J A S O N D

Mean annual rainfall : 749 mm
Mean January temperature : 12.5°
Mean July temperature : 17.5°C

Oblique Mercator Projection
© Oxford University Press

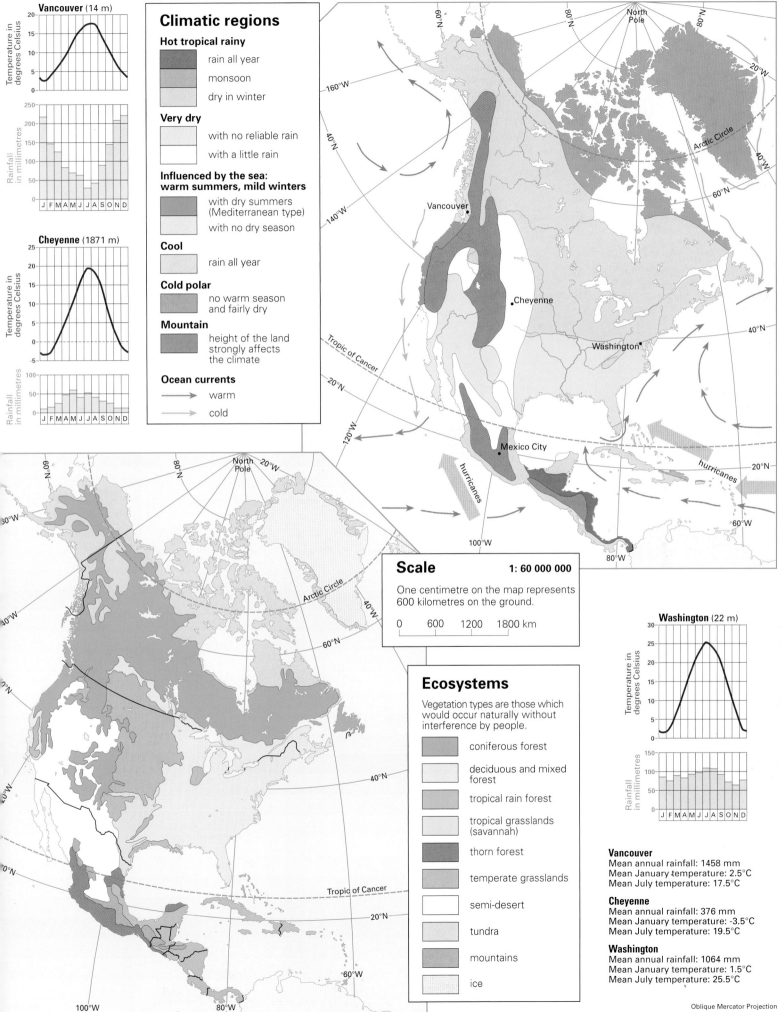

Vancouver (14 m)

Temperature in degrees Celsius

Rainfall in millimetres

J F M A M J J A S O N D

Cheyenne (1871 m)

Temperature in degrees Celsius

Rainfall in millimetres

J F M A M J J A S O N D

Climatic regions

Hot tropical rainy
- rain all year
- monsoon
- dry in winter

Very dry
- with no reliable rain
- with a little rain

Influenced by the sea: warm summers, mild winters
- with dry summers (Mediterranean type)
- with no dry season

Cool
- rain all year

Cold polar
- no warm season and fairly dry

Mountain
- height of the land strongly affects the climate

Ocean currents
- → warm
- → cold

Scale

1: 60 000 000

One centimetre on the map represents 600 kilometres on the ground.

0   600   1200   1800 km

Ecosystems

Vegetation types are those which would occur naturally without interference by people.

- coniferous forest
- deciduous and mixed forest
- tropical rain forest
- tropical grasslands (savannah)
- thorn forest
- temperate grasslands
- semi-desert
- tundra
- mountains
- ice

Washington (22 m)

Temperature in degrees Celsius

Rainfall in millimetres

J F M A M J J A S O N D

Vancouver
Mean annual rainfall: 1458 mm
Mean January temperature: 2.5°C
Mean July temperature: 17.5°C

Cheyenne
Mean annual rainfall: 376 mm
Mean January temperature: -3.5°C
Mean July temperature: 19.5°C

Washington
Mean annual rainfall: 1064 mm
Mean January temperature: 1.5°C
Mean July temperature: 25.5°C

Oblique Mercator Projection
© Oxford University Press

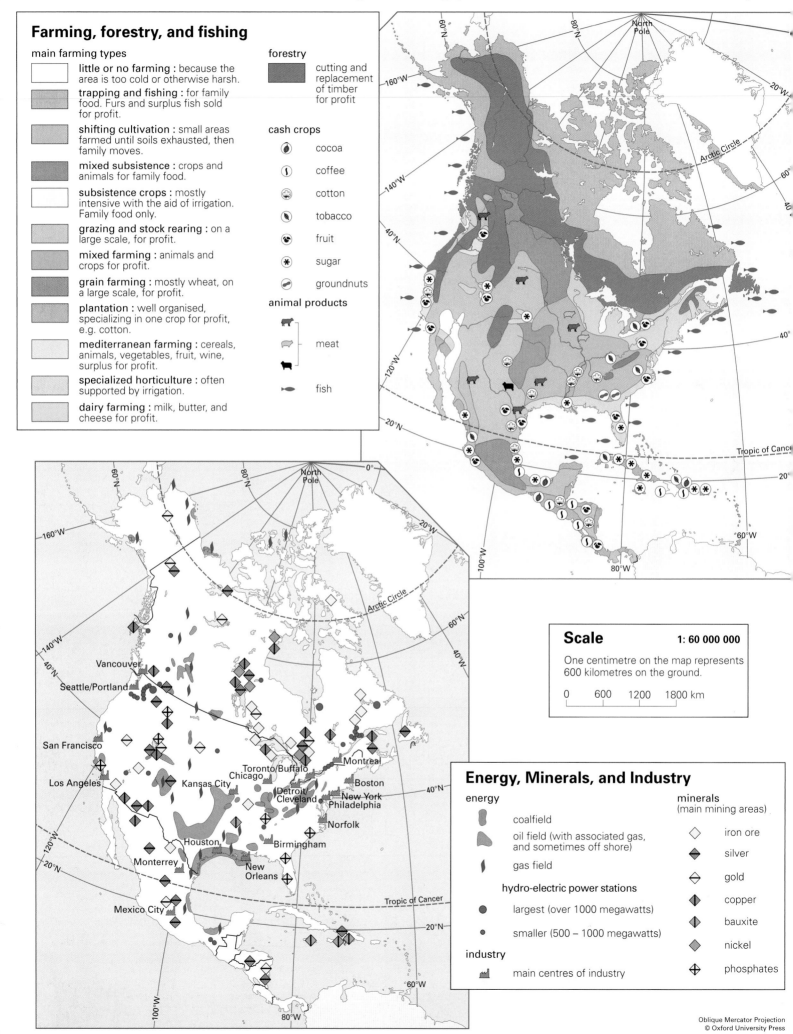

Farming, forestry, and fishing

main farming types

little or no farming : because the area is too cold or otherwise harsh.

trapping and fishing : for family food. Furs and surplus fish sold for profit.

shifting cultivation : small areas farmed until soils exhausted, then family moves.

mixed subsistence : crops and animals for family food.

subsistence crops : mostly intensive with the aid of irrigation. Family food only.

grazing and stock rearing : on a large scale, for profit.

mixed farming : animals and crops for profit.

grain farming : mostly wheat, on a large scale, for profit.

plantation : well organised, specializing in one crop for profit, e.g. cotton.

mediterranean farming : cereals, animals, vegetables, fruit, wine, surplus for profit.

specialized horticulture : often supported by irrigation.

dairy farming : milk, butter, and cheese for profit.

forestry

cutting and replacement of timber for profit

cash crops

cocoa

coffee

cotton

tobacco

fruit

sugar

groundnuts

animal products

meat

fish

Scale   1: 60 000 000

One centimetre on the map represents 600 kilometres on the ground.

0   600   1200   1800 km

Energy, Minerals, and Industry

energy

coalfield

oil field (with associated gas, and sometimes off shore)

gas field

hydro-electric power stations

largest (over 1000 megawatts)

smaller (500 – 1000 megawatts)

industry

main centres of industry

minerals
(main mining areas)

iron ore

silver

gold

copper

bauxite

nickel

phosphates

Oblique Mercator Projection
© Oxford University Press

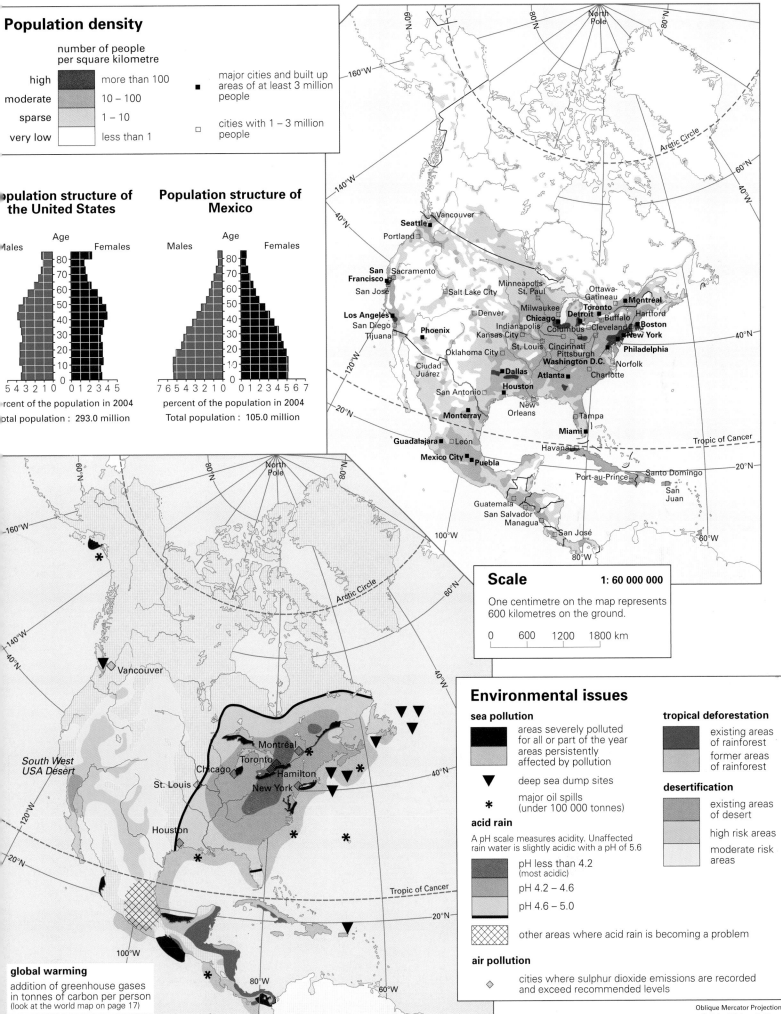

Population density

number of people
per square kilometre

high	more than 100
moderate	10 – 100
sparse	1 – 10
very low	less than 1

■ major cities and built up areas of at least 3 million people

□ cities with 1 – 3 million people

Population structure of the United States

Age

Males   Females

5 4 3 2 1 0   0 1 2 3 4 5
percent of the population in 2004
Total population : 293.0 million

Population structure of Mexico

Age

Males   Females

7 6 5 4 3 2 1 0   0 1 2 3 4 5 6 7
percent of the population in 2004
Total population : 105.0 million

Scale   1: 60 000 000

One centimetre on the map represents 600 kilometres on the ground.

0   600   1200   1800 km

Environmental issues

sea pollution

■ areas severely polluted for all or part of the year

areas persistently affected by pollution

▼ deep sea dump sites

✴ major oil spills (under 100 000 tonnes)

acid rain

A pH scale measures acidity. Unaffected rain water is slightly acidic with a pH of 5.6

pH less than 4.2 (most acidic)

pH 4.2 – 4.6

pH 4.6 – 5.0

other areas where acid rain is becoming a problem

air pollution

◇ cities where sulphur dioxide emissions are recorded and exceed recommended levels

tropical deforestation

existing areas of rainforest

former areas of rainforest

desertification

existing areas of desert

high risk areas

moderate risk areas

global warming

addition of greenhouse gases in tonnes of carbon per person
(look at the world map on page 17)

South West USA Desert

Oblique Mercator Projection
© Oxford University Press

Key

| international boundary |
| state or province boundary |
| motorway and main road |
| railway |
| canal |

peak or highest point

towns
- largest
- large
- others

major airport
river and dam
lake
ice cap
marsh

Land height

in metres above sea level
- more than 2000 m
- 1000 – 2000 m
- 500 – 1000 m
- 200 – 500m
- less than 200 m
- below sea level

Scale

1: 25 000 000

One centimetre on the map measures 250 kilometres on the ground.

0 250 500 750 1000 km

RUSSIAN FEDERATION (RUSSIA)

GREENLAND

ICELAND

Reykjavík

Mt. Forel 3360 m

Arctic Circle

Cape Farewell

Atlantic Ocean

NEWFOUNDLAND AND LABRADOR

Nuuk (Godthåb)

Smallwood Reservoir

Churchill Falls

Schefferville

La Grande Rivière

Baffin Bay

Baffin Island

Southampton Island

Hudson Bay

CANADA

Churchill

Nelson

Lynn Lake

MANITOBA

Lake Winnipeg

Ellesmere Island

Devon Island

Sverdrup Islands

Queen Elizabeth Islands

Parry Islands

Melville Island

Victoria Island

Banks Island

NUNAVUT

Great Bear Lake

Great Slave Lake

Yellowknife

Hay River

NORTHWEST TERRITORIES

Mackenzie

Fort Simpson

Liard

MACKENZIE MOUNTAINS

Peace

Athabasca

Fort McMurray

SASKATCHEWAN

Saskatchewan

Saskatoon

Regina

ALBERTA

Edmonton

Calgary

Mt. Robson 3954 m

BRITISH COLUMBIA

ROCKY MOUNTAINS

Columbia

Fraser

Prince Rupert

COAST MOUNTAINS

Mt. Waddington 4042 m

Queen Charlotte Islands

Vancouver Island

Victoria

Vancouver

Seattle

Tacoma

WASHINGTON

Spokane

Portland

Mt. Rainier 4392 m

Arctic Ocean

Beaufort Sea

Prudhoe Bay

Inuvik

Dawson

YUKON TERRITORY

Whitehorse

Mt. Logan 5951 m

BROOKS RANGE

ALASKA

Fairbanks

Yukon

ALASKA RANGE

Mt. McKinley 6194 m

Anchorage

Seward

Gulf of Alaska

Kodiak Island

Pacific Ocean

Bering Strait

St. Lawrence

Bering Sea

St. Matthew

Nunivak

Alaska Peninsula

Unimak Island

North Pole

Arctic Circle

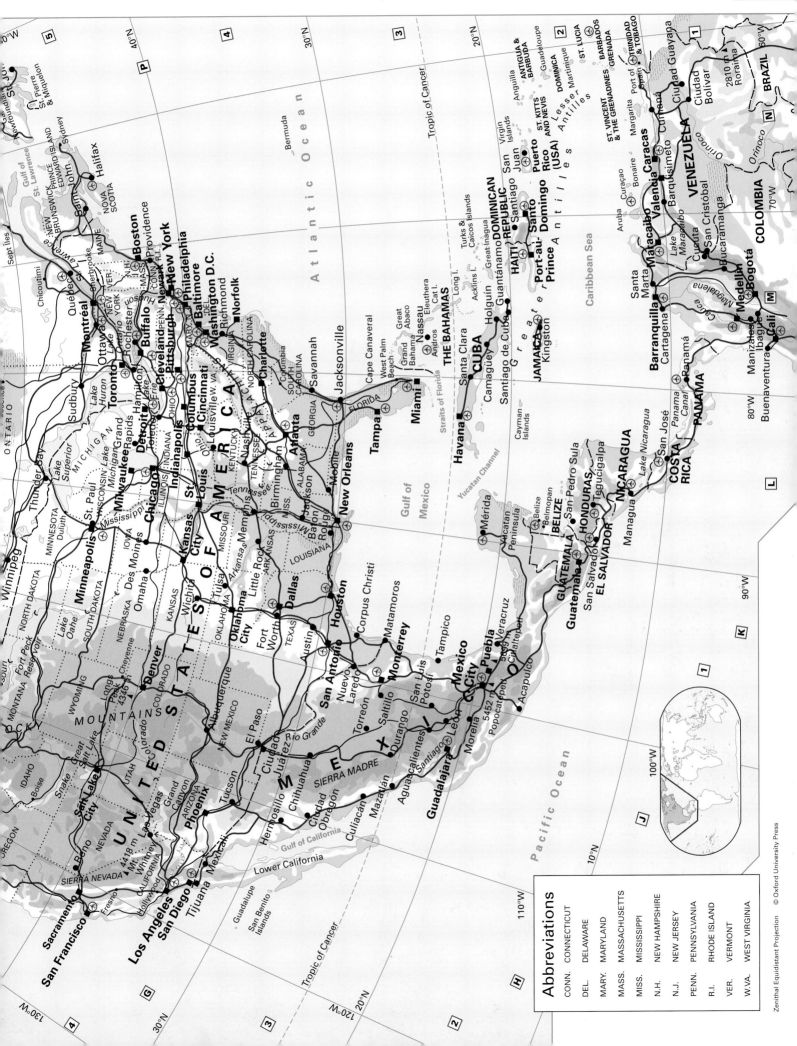

Abbreviations

CONN.	CONNECTICUT
DEL.	DELAWARE
MARY.	MARYLAND
MASS.	MASSACHUSETTS
MISS.	MISSISSIPPI
N.H.	NEW HAMPSHIRE
N.J.	NEW JERSEY
PENN.	PENNSYLVANIA
R.I.	RHODE ISLAND
VER.	VERMONT
W.VA.	WEST VIRGINIA

Countries and capitals

country boundary

• capital city

The British Isles at the same scale

100°W · 80°W · 60°W · 40°W

Aruba *(Neths.)* · *Netherlands Antilles (Netherlands)*

Caracas
VENEZUELA · Georgetown
Paramaribo
Cayenne
French Guiana (France)
• Bogotá · **GUYANA** · **SURINAME**
COLOMBIA

0° Equator · Equator

Galapagos Islands (Ecuador) · Quito
ECUADOR

PERU · **BRAZIL**

Lima

La Paz · • Brasília
BOLIVIA

PARAGUAY

20°S · Tropic of Capricorn · Asunción · Tropic of Capricorn · 20°

URUGUAY

Santiago · **ARGENTINA** · Buenos Aires · Montevideo

CHILE

40°S · 40°

Stanley
Falkland Islands (UK)

· *South Georgia (UK)*

South Shetland Islands (UK) · *South Orkney Islands (UK)* · 20°S
80°W · 60°W · 40°W · 60°S
Prime Meridian

Scale · **1: 60 000 000**

One centimetre on the map represents 600 kilometres on the ground.

0 · 600 · 1200 · 1800 km

100°W · 80°W · 60°W · 40°W

Caribbean Sea

Lake Maracaibo
Angel Falls
Magdalena · Llanos · Orinoco · **GUIANA HIGHLANDS**

Equator · 0° · Equator · 0°

▲5896 Cotopaxi
6310 Chimborazo · Negro · Amazon
Galapagos Islands · *Selvas* · *Tapajós*
Madeira · *Xingu* · *Tocantins* · **BRAZILIAN**
São Francisco
MATO GROSSO · **HIGHLANDS**
Lake Titicaca
Lake Poopó · **BRAZIL PLATEAU**

PACIFIC OCEAN

Pilcomayo · *Paraguay*

20°S · *Atacama Desert* · 20°S · 20°S
Tropic of Capricorn · Gran Chaco · *Paraná* · Tropic of Capricorn
▲6908 · *Uruguay*
Ojos del Salado
6960▲ · *Salado* · River Plate Estuary
Aconcagua · *Colorado* · Pampas · **ATLANTIC OCEAN**

ANDES · *Negro*
Valdes Peninsula *(40m below sea level)*
Patagonia

40°S · Falkland Islands · 40°S
Tierra del Fuego · South Georgia
Cape Horn
SOUTHERN · **OCEAN**
100°W · 80°W · 60°W · Drake Passage · 20°S
60°W · 40°W · 60°S · *Prime Meridian*

Land height

in metres above sea level

more than 5000 m

2000 – 5000 m

1000 – 2000 m

500 – 1000 m

200 – 500 m

sea level – 200 m

below sea level

▲ highest peaks with heights in metres

lakes

major rivers

marsh

ice cap

Oblique Mercator Proj
© Oxford University

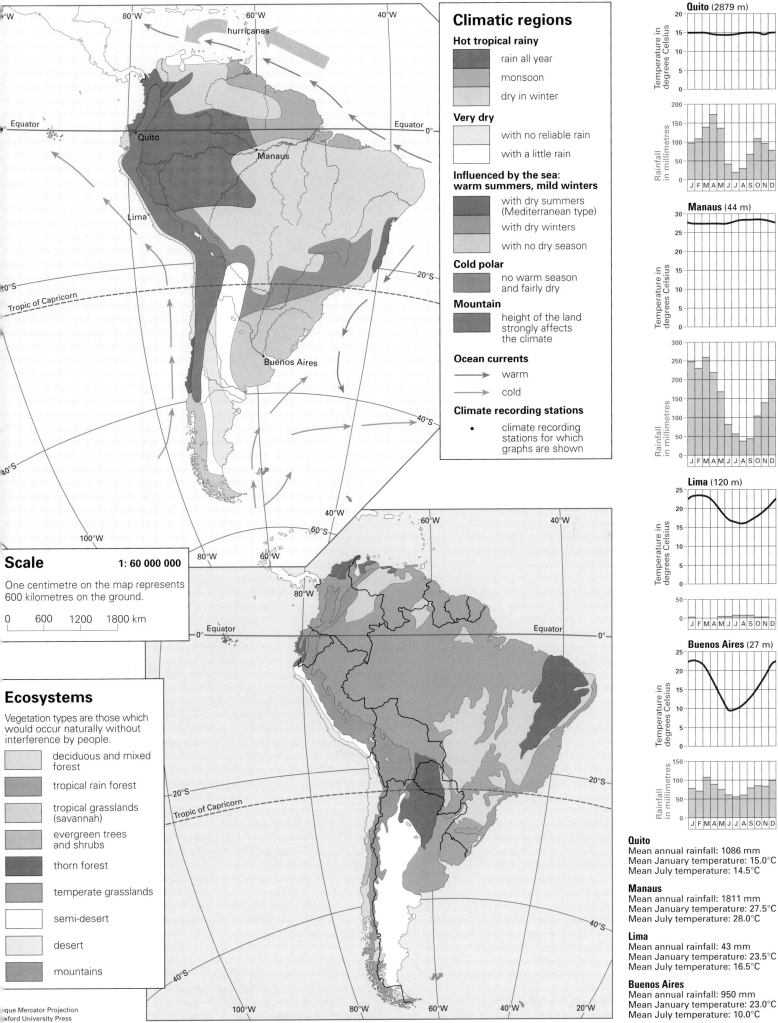

Climatic regions

Hot tropical rainy
- rain all year
- monsoon
- dry in winter

Very dry
- with no reliable rain
- with a little rain

Influenced by the sea: warm summers, mild winters
- with dry summers (Mediterranean type)
- with dry winters
- with no dry season

Cold polar
- no warm season and fairly dry

Mountain
- height of the land strongly affects the climate

Ocean currents
- → warm
- → cold

Climate recording stations
- • climate recording stations for which graphs are shown

Scale 1: 60 000 000

One centimetre on the map represents 600 kilometres on the ground.

0   600   1200   1800 km

Ecosystems

Vegetation types are those which would occur naturally without interference by people.

- deciduous and mixed forest
- tropical rain forest
- tropical grasslands (savannah)
- evergreen trees and shrubs
- thorn forest
- temperate grasslands
- semi-desert
- desert
- mountains

que Mercator Projection
xford University Press

Quito (2879 m)

Temperature in degrees Celsius

Rainfall in millimetres
J F M A M J J A S O N D

Manaus (44 m)

Temperature in degrees Celsius

Rainfall in millimetres
J F M A M J J A S O N D

Lima (120 m)

Temperature in degrees Celsius

Rainfall in millimetres
J F M A M J J A S O N D

Buenos Aires (27 m)

Temperature in degrees Celsius

Rainfall in millimetres
J F M A M J J A S O N D

Quito
Mean annual rainfall: 1086 mm
Mean January temperature: 15.0°C
Mean July temperature: 14.5°C

Manaus
Mean annual rainfall: 1811 mm
Mean January temperature: 27.5°C
Mean July temperature: 28.0°C

Lima
Mean annual rainfall: 43 mm
Mean January temperature: 23.5°C
Mean July temperature: 16.5°C

Buenos Aires
Mean annual rainfall: 950 mm
Mean January temperature: 23.0°C
Mean July temperature: 10.0°C

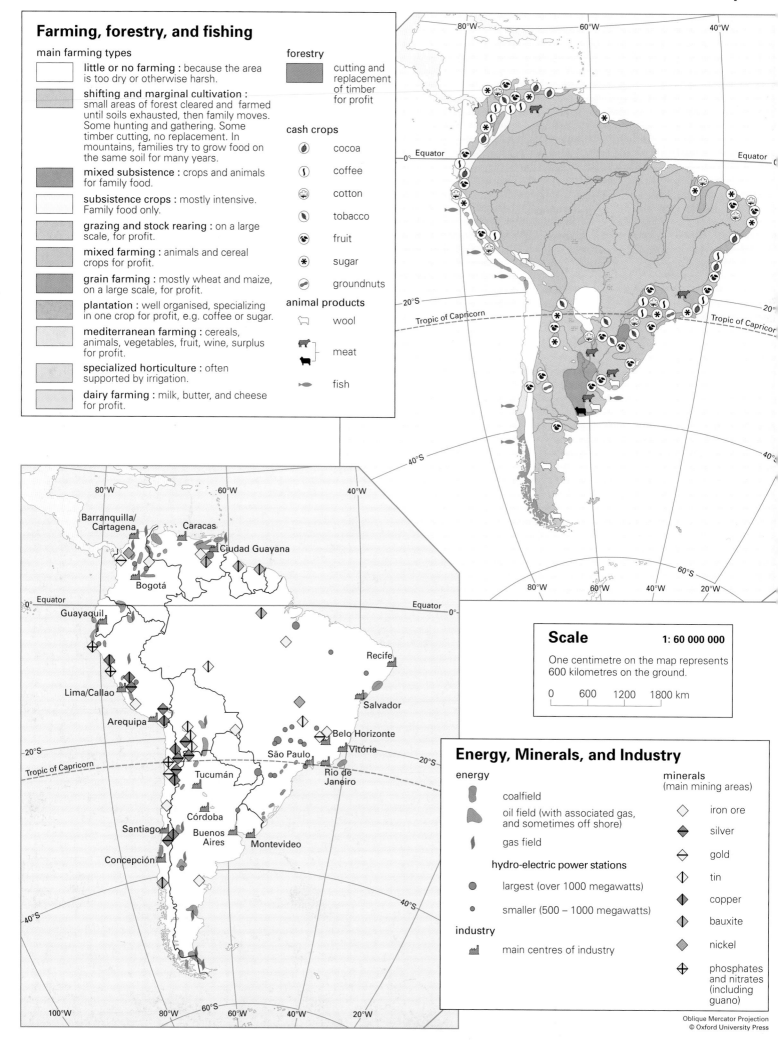

Farming, forestry, and fishing

main farming types

little or no farming : because the area is too dry or otherwise harsh.

shifting and marginal cultivation : small areas of forest cleared and farmed until soils exhausted, then family moves. Some hunting and gathering. Some timber cutting, no replacement. In mountains, families try to grow food on the same soil for many years.

mixed subsistence : crops and animals for family food.

subsistence crops : mostly intensive. Family food only.

grazing and stock rearing : on a large scale, for profit.

mixed farming : animals and cereal crops for profit.

grain farming : mostly wheat and maize, on a large scale, for profit.

plantation : well organised, specializing in one crop for profit, e.g. coffee or sugar.

mediterranean farming : cereals, animals, vegetables, fruit, wine, surplus for profit.

specialized horticulture : often supported by irrigation.

dairy farming : milk, butter, and cheese for profit.

forestry

cutting and replacement of timber for profit

cash crops

- cocoa
- coffee
- cotton
- tobacco
- fruit
- sugar
- groundnuts

animal products

- wool
- meat
- fish

Scale

1: 60 000 000

One centimetre on the map represents 600 kilometres on the ground.

0   600   1200   1800 km

Energy, Minerals, and Industry

energy

coalfield

oil field (with associated gas, and sometimes off shore)

gas field

hydro-electric power stations

largest (over 1000 megawatts)

smaller (500 – 1000 megawatts)

industry

main centres of industry

minerals
(main mining areas)

- iron ore
- silver
- gold
- tin
- copper
- bauxite
- nickel
- phosphates and nitrates (including guano)

Oblique Mercator Projection
© Oxford University Press

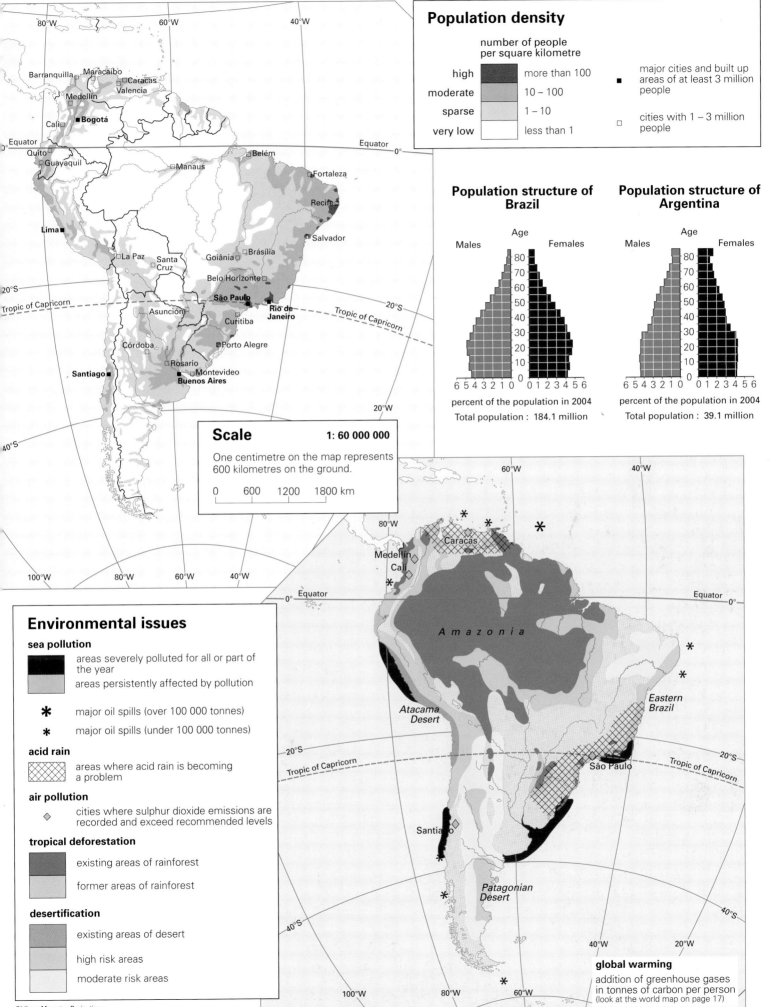

Population density

number of people per square kilometre

high		more than 100
moderate		10 – 100
sparse		1 – 10
very low		less than 1

■ major cities and built up areas of at least 3 million people

□ cities with 1 – 3 million people

Population structure of Brazil

Age

Males — Females

80
70
60
50
40
30
20
10
0

6 5 4 3 2 1 0   0 1 2 3 4 5 6

percent of the population in 2004

Total population : 184.1 million

Population structure of Argentina

Age

Males — Females

80
70
60
50
40
30
20
10
0

6 5 4 3 2 1 0   0 1 2 3 4 5 6

percent of the population in 2004

Total population : 39.1 million

Scale

1: 60 000 000

One centimetre on the map represents 600 kilometres on the ground.

0   600   1200   1800 km

Environmental issues

sea pollution

areas severely polluted for all or part of the year

areas persistently affected by pollution

✳ major oil spills (over 100 000 tonnes)

✳ major oil spills (under 100 000 tonnes)

acid rain

areas where acid rain is becoming a problem

air pollution

◇ cities where sulphur dioxide emissions are recorded and exceed recommended levels

tropical deforestation

existing areas of rainforest

former areas of rainforest

desertification

existing areas of desert

high risk areas

moderate risk areas

global warming

addition of greenhouse gases in tonnes of carbon per person (look at the world map on page 17)

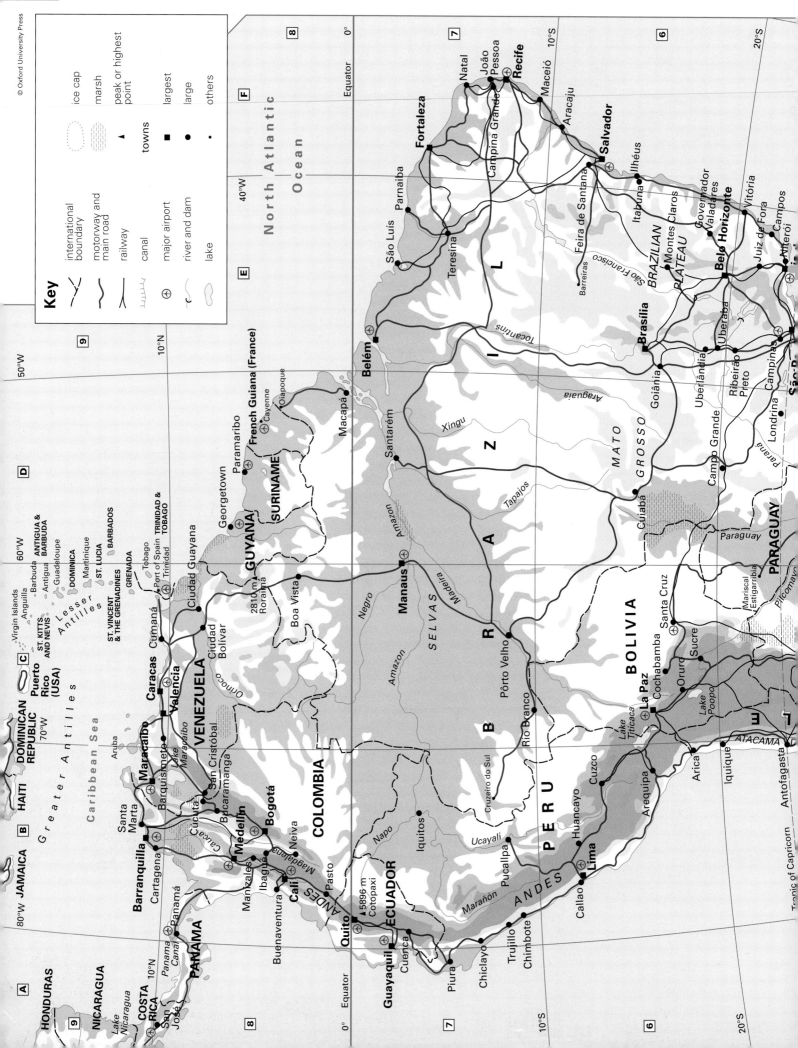

Key

international boundary		ice cap
motorway and main road		marsh
railway		peak or highest point
canal		towns
major airport		largest
river and dam		large
lake		others

North Atlantic Ocean

Caribbean Sea

North Pacific Ocean

HONDURAS
NICARAGUA
COSTA RICA
Lake Nicaragua
San José
PANAMA
Panamá
Panama Canal

JAMAICA
HAITI
DOMINICAN REPUBLIC
Puerto Rico (USA)
Virgin Islands
Anguilla
ANTIGUA & BARBUDA
Barbuda
Antigua
Guadeloupe
DOMINICA
Martinique
ST. LUCIA
BARBADOS
ST. VINCENT & THE GRENADINES
GRENADA
Lesser Antilles
Greater Antilles
Aruba
Tobago
TRINIDAD & TOBAGO
Port of Spain
Trinidad

COLOMBIA
Santa Marta
Barranquilla
Cartagena
Maracaibo
Lake Maracaibo
Barquisimeto
Valencia
Caracas
Cumaná
VENEZUELA
San Cristóbal
Cúcuta
Bucaramanga
Medellín
Manizales
Ibagué
Bogotá
Cali
Neiva
Pasto
Buenaventura
Cauca
Magdalena
Orinoco
Ciudad Bolívar
Ciudad Guayana
GUYANA
Georgetown
SURINAME
Paramaribo
French Guiana (France)
Cayenne
Oiapoque
Roraima 2810 m
Boa Vista

ECUADOR
Quito
Cotopaxi 5896 m
Guayaquil
Cuenca
Piura
Chiclayo
Trujillo
Chimbote
Callao
Lima
PERU
Iquitos
Napo
Marañón
Ucayali
Pucallpa
Huancayo
Cuzco
ANDES
Arequipa
Arica
Iquique
Antofagasta
Tropic of Capricorn
ATACAMA

Belém
Macapá
Amazon
Santarém
Xingu
Tapajos
Manaus
Negro
Madeira
SELVAS
B R A Z I L
Pôrto Velho
Rio Branco
Cruzeiro do Sul
BOLIVIA
La Paz
Lake Titicaca
Cochabamba
Oruro
Sucre
Santa Cruz
Lake Poopó
Mariscal Estigarribia
PARAGUAY
Pilcomayo
Paraguay

Natal
João Pessoa
Campina Grande
Recife
Maceió
Aracaju
Fortaleza
Parnaíba
São Luís
Teresina
Feira de Santana
Salvador
Itabuna
Ilhéus
Barreiras
São Francisco
Montes Claros
Governador Valadares
BRAZILIAN PLATEAU
Brasília
Goiânia
Belo Horizonte
Uberaba
Uberlândia
Ribeirão Preto
Juiz de Fora
Vitória
Campos
Niterói
Campinas
Londrina
Tocantins
Araguaia
MATO GROSSO
Cuiabá
Campo Grande
Paraná

Equator

50°W 60°W 70°W 80°W 40°W 10°N 10°S 20°S 0°

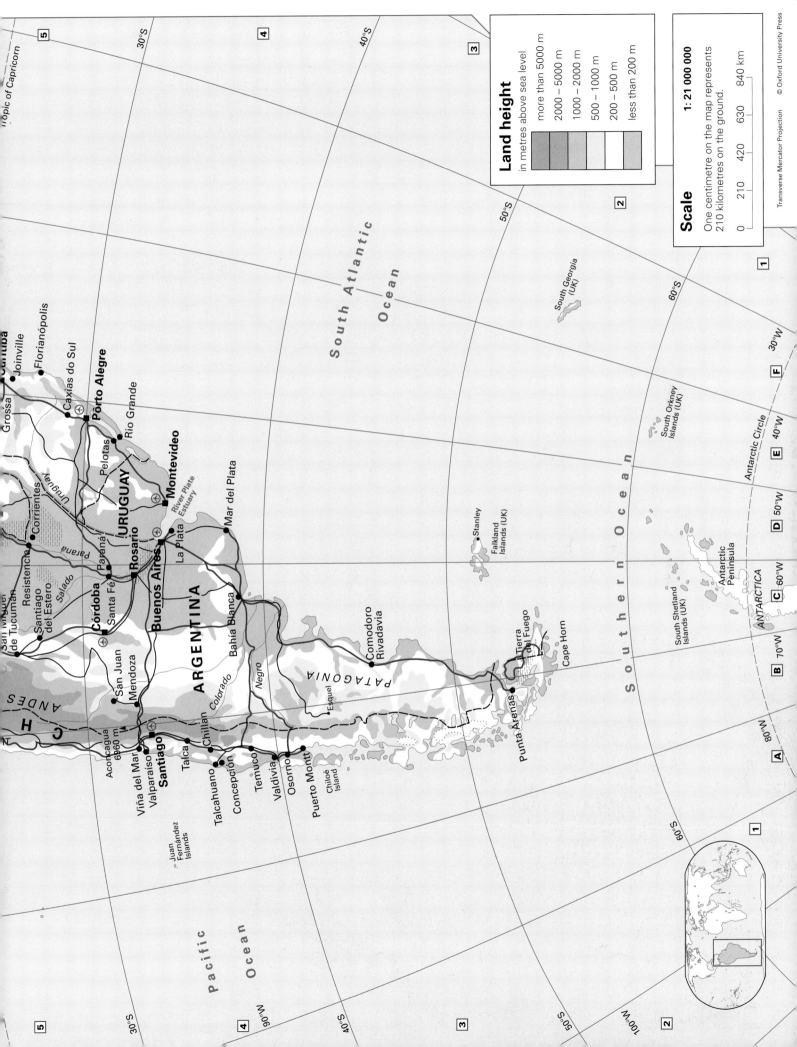

Transverse Mercator Projection © Oxford University Press

Land height

in metres above sea level

	more than 5000 m
	2000 – 5000 m
	1000 – 2000 m
	500 – 1000 m
	200 – 500 m
	less than 200 m

Scale

1 : 21 000 000

One centimetre on the map represents 210 kilometres on the ground.

0	210	420	630	840 km

Tropic of Capricorn

Curitiba
Joinville
Florianópolis
Caxias do Sul
Porto Alegre
Rio Grande
Pelotas
Montevideo
River Plate Estuary
Mar del Plata
La Plata
Buenos Aires
Rosario
Paraná
Santa Fé
Córdoba
Santiago del Estero
San Miguel de Tucumán
Resistencia
Corrientes
URUGUAY
Uruguay
Paraná
Salado
ARGENTINA
San Juan
Mendoza
Aconcagua 6960 m
Viña del Mar
Valparaíso
Santiago
Talca
Chillán
Talcahuano
Concepción
Temuco
Valdivia
Osorno
Puerto Montt
Chiloé Island
Esquel
Bahía Blanca
Colorado
Negro
Comodoro Rivadavia
PATAGONIA
ANDES
CHILE
Juan Fernández Islands

South Atlantic Ocean

South Georgia (UK)

Stanley
Falkland Islands (UK)

South Orkney Islands (UK)

South Shetland Islands (UK)

Antarctic Peninsula

ANTARCTICA

Southern Ocean

Punta Arenas
Tierra del Fuego
Cape Horn

Pacific Ocean

30°S
40°S
50°S
60°S

80°W
70°W
60°W
50°W
40°W
30°W

90°W
100°W

Antarctic Circle

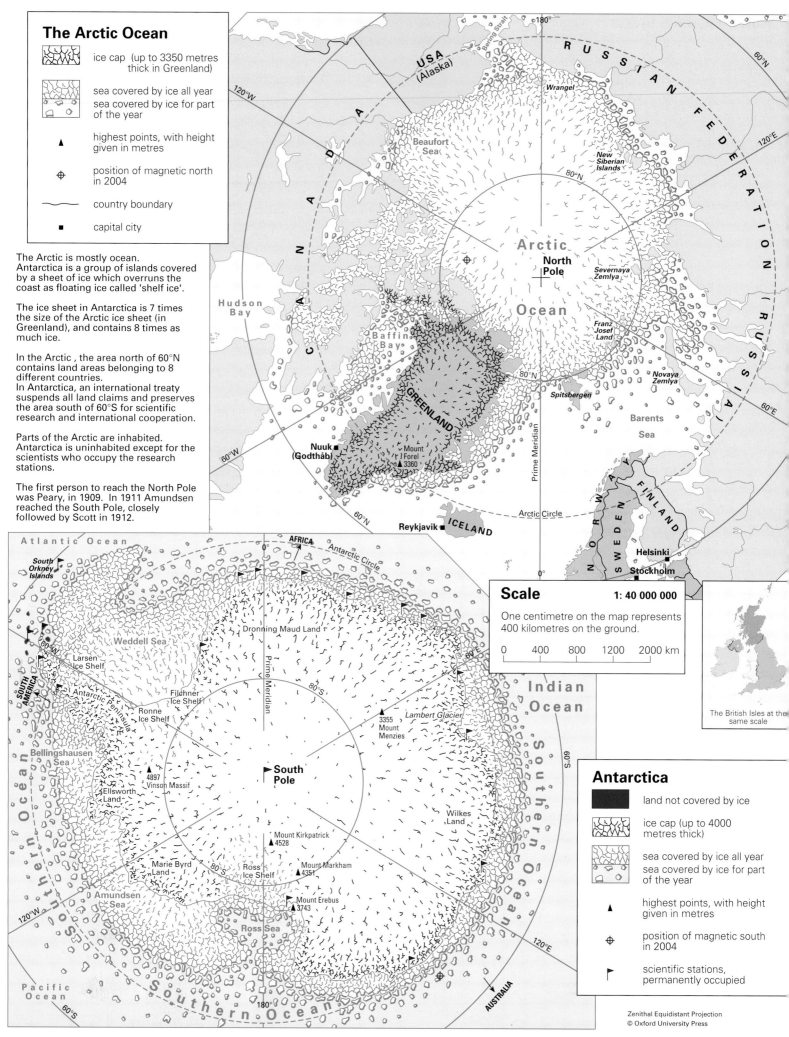

The Arctic Ocean

- ice cap (up to 3350 metres thick in Greenland)
- sea covered by ice all year
- sea covered by ice for part of the year
- ▲ highest points, with height given in metres
- ⊕ position of magnetic north in 2004
- country boundary
- ■ capital city

The Arctic is mostly ocean.
Antarctica is a group of islands covered by a sheet of ice which overruns the coast as floating ice called 'shelf ice'.

The ice sheet in Antarctica is 7 times the size of the Arctic ice sheet (in Greenland), and contains 8 times as much ice.

In the Arctic , the area north of 60°N contains land areas belonging to 8 different countries.
In Antarctica, an international treaty suspends all land claims and preserves the area south of 60°S for scientific research and international cooperation.

Parts of the Arctic are inhabited.
Antarctica is uninhabited except for the scientists who occupy the research stations.

The first person to reach the North Pole was Peary, in 1909. In 1911 Amundsen reached the South Pole, closely followed by Scott in 1912.

Scale

1: 40 000 000

One centimetre on the map represents 400 kilometres on the ground.

0   400   800   1200   2000 km

The British Isles at the same scale

Antarctica

- land not covered by ice
- ice cap (up to 4000 metres thick)
- sea covered by ice all year
- sea covered by ice for part of the year
- ▲ highest points, with height given in metres
- ⊕ position of magnetic south in 2004
- ⚑ scientific stations, permanently occupied

Zenithal Equidistant Projection
© Oxford University Press

How to use the index

To find a place on an atlas map use either the grid code or latitude and longitude.

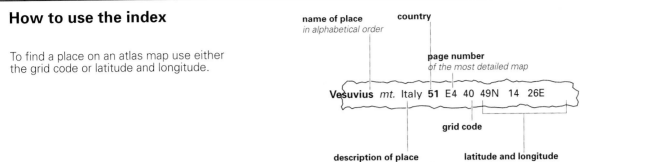

name of place
in alphabetical order

country

page number
of the most detailed map

Vesuvius *mt.* Italy **51** E4 40 49N 14 26E

description of place
(see list of abbreviations)

grid code

latitude and longitude
sometimes approximate

Grid code

Vesuvius is in grid square E4

Vesuvius *mt.* Italy **51** E4 40 49N 14 26E

Latitude and longitude

Vesuvius is at latitude 40 49N longitude 14 26E

Vesuvius *mt.* Italy **51** E4 40 49N 14 26E

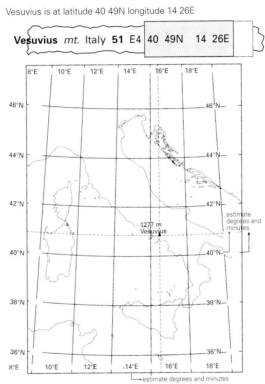

Abbreviations used in the index

admin.	administrative area
b.	bay or harbour
bor.	borough
c.	cape, point or headland
co.	county
est.	estuary
geog.reg.	geographical region
i.	island
is.	islands
l.	lake, lakes, lagoon
mt.	mountain
mts.	mountains
p.	peninsula
pk.	peak
plat.	plateau
pt.	point
r.	river
res.	reservoir
sd.	sound, strait or channel
sum.	summit
tn.	town
u.a.	unitary authority
vol.	volcano

A

Aachen Germany **49** D4 50 46N 6 06E
Aare *r.* Switzerland **49** D2 47 15N 7 30E
Abadan Iran **61** E4 30 20N 48 15E
Abbeville France **48** D5 50 06N 1 51E
Aberaeron Wales **36** C2 52 49N 44 43W
Aberchirder Scotland **31** G2 57 33N 2 38W
Aberdare Wales **36** D1 51 43N 3 27W
Aberdeen Scotland **31** G2 57 10N 2 04W
Aberdeen City *u.a.* Scotland **31** G2 57 10N 2 00W
Aberdeenshire *u.a.* Scotland **31** G2 57 10N 2 50W
Aberfeldy Scotland **31** F1 56 37N 3 54W
Abergavenny Wales **36** D1 51 50N 3 00W
Abertillery Wales **36** D1 51 45N 3 09W
Aberystwyth Wales **36** C2 52 25N 4 05W
Abha Saudi Arabia **61** E2 18 14N 42 31E
Abidjan Côte d'Ivoire **66** B3 5 19N 4 01W
Abingdon England **38** C2 51 41N 1 17W
Aboyne Scotland **31** G2 57 05N 2 50W
Abu Dhabi United Arab Emirates **61** F3 24 28N 54 25E
Abuja Nigeria **66** C3 9 10N 7 11E
Acapulco Mexico **79** K2 16 51N 99 56W
Accra Ghana **66** B3 5 33N 0 15W
Acklins Island The Bahamas **79** M3 22 30N 74 30W
Aconcagua *mt.* Argentina **85** B4 32 40S 70 02W
A Coruña Spain **50** A5 43 22N 8 24W
Adamawa Highlands Africa **66** D3 7 00N 13 00E
Adana Turkey **43** P3 37 00N 35 19E
Addis Ababa Ethiopia **61** D1 9 03N 38 42E
Adelaide Australia **72** D2 34 55S 138 36E
Aden Yemen Republic **61** E2 12 50N 45 03E
Aden, Gulf of Indian Ocean **61** E2 12 30N 47 30E
Adour *r.* France **48** C1 43 45N 0 30W
Adriatic Sea Mediterranean Sea **51** E5 43 00N 15 00E
Aegean Sea Mediterranean Sea **42** L3 39 00N 24 00E
AFGHANISTAN 58 B4
Agadez Niger **66** C4 17 00N 7 56E
Agadir Morocco **66** B6 30 30N 9 40W
Agra India **58** C3 27 09N 78 00E
Aguascalientes Mexico **79** J3 21 51N 102 18W
Ahmadabad India **58** C3 23 03N 72 40E
Ahvaz Iran **61** E4 31 17N 48 43E

Ailsa Craig *i.* Scotland **32** D3 55 16N 5 07W
Aïn Sefra Algeria **42** F2 32 45N 0 35W
Airdrie Scotland **33** F3 55 52N 3 59W
Aire *r.* England **34** C2 54 00N 2 05W
Aix-en-Provence France **48** F1 43 31N 5 27E
Ajaccio Corsica **51** B4 41 55N 8 43E
Ajdabiya Libya **66** E6 30 46N 20 14E
Akita Japan **59** N2 39 44N 140 05E
Akureyri Iceland **42** C9 65 41N 18 04W
Alabama *state* USA **79** L4 32 00N 87 00W
Alaska *state* USA **78** D7 63 00N 150 00W
Alaska, Gulf of USA **78** E6 58 00N 147 00W
Alaska Peninsula USA **78** D6 56 30N 159 00W
Alaska Range *mts.* USA **78** D7/E7 62 30N 152 30W
Albacete Spain **50** E3 39 00N 1 52W
ALBANIA 42 L4
Albany Australia **72** B2 35 00S 117 53E
Alberta *province* Canada **78** H6 55 00N 115 00W
Albert, Lake Congo Dem. Rep./Uganda **67** D5 2 00N 31 00E
Ålborg Denmark **42** H7 57 05N 9 50E
Albuquerque USA **79** J4 35 05N 106 38W
Alcalá de Henares Spain **50** D4 40 28N 3 22W
Alcudia Balearic Islands **50** G3 39 51N 3 06E
Aldabra Islands Indian Ocean **67** E4 9 00S 46 00E
Aldeburgh England **39** F3 52 09N 1 35E
Alderney *i.* Channel Islands British Isles **41** E2 49 43N 2 12W
Aldershot England **38** D2 51 15N 0 47W
Aleppo Syria **61** D4 36 14N 37 10E
Alessándria Italy **51** B6 44 55N 8 37E
Alexandria Egypt **61** C4 31 13N 29 55E
Alexandria Scotland **33** E3 55 59N 4 36W
Algarve *geog. reg.* Portugal **50** A2 37 30N 8 00W
ALGERIA 66 C5
Algiers Algeria **66** C6 36 50N 3 00E
Al Hoceima Morocco **50** D1 35 14N 3 56W
Alicante Spain **50** E3 38 21N 0 29W
Alice Springs Australia **72** D3 23 41S 133 52E
Al Jawf Libya **66** E5 24 12N 23 18E
Allahabad India **58** D3 25 27N 81 50E
Allier *r.* France **48** E3 46 15N 3 15E
Alloa Scotland **33** F4 56 07N 3 49W
Almanzor *mt.* Spain **50** C4 40 15N 5 18W

Almaty Kazakhstan **56** H2 43 19N 76 55E
Almería Spain **50** D2 36 50N 2 26W
Al Mukha Yemen Republic **61** E2 13 20N 43 16E
Aln *r.* England **33** H3 55 30N 1 50W
Alnwick England **33** H3 55 25N 1 42W
Alps *mts.* Europe **49** D2/G2 46 00N 7 30E
Altai Mountains Mongolia **57** K2 47 00N 92 30E
Alton England **38** D2 51 09N 0 59W
Alyth Scotland **31** F1 56 37N 3 13W
Amazon *r.* Brazil **84** D7 2 30S 65 30W
Amble England **33** H3 55 20N 1 34W
Ambleside England **34** C3 54 26N 2 58W
Ambon Indonesia **60** D2 3 41S 128 10E
Amesbury England **38** C2 51 10N 1 47W
Amiens France **48** E4 49 54N 2 18E
Amlwch Wales **36** C3 53 25N 4 20W
Amman Jordan **61** D4 31 04N 46 17E
Ammanford Wales **36** D1 51 48N 3 58W
Amritsar India **58** C4 31 35N 74 56E
Amsterdam Netherlands **49** C5 52 22N 4 54E
Amu Darya *r.* Asia **56** G2 41 00N 61 00E
Amundsen Sea Southern Ocean **86** 72 00S 130 00W
Amur *r.* Asia **57** N3 54 00N 122 00E
Anchorage USA **78** E7 61 10N 150 00W
Ancona Italy **51** D5 43 37N 13 31E
Andaman Islands India **58** E2 12 00N 94 00E
Andaman Sea Indian Ocean **58** E2 13 00N 95 00E
Andes *mts.* South America **84/85** B8/C5 10 00S 77 00W
Andizhan Uzbekistan **56** H2 40 40N 72 12E
ANDORRA 50 F5
Andover England **38** C2 51 13N 1 28W
Andros *i.* The Bahamas **79** M3 24 00N 78 00W
Aneto *mt.* Spain **50** F5 42 37N 0 40E
Angara *r.* Russia **57** K3 59 00N 97 00E
Angeles The Philippines **60** D4 15 09N 120 33E
Angers France **48** C3 47 29N 0 32W
Anglesey *i.* Wales **36** C3 53 13N 4 23W
ANGOLA 67 B3
Angoulême France **48** D2 45 40N 0 10E
Anguilla *i.* Leeward Islands **79** N2 18 14N 63 05W
Angus *u.a.* Scotland **31** F1/G1 56 45N 3 00W
Ankara Turkey **43** N3 39 55N 32 50E
'Annaba Algeria **66** C6 36 55N 7 47E

An Najaf Iraq **61** E4 31 59N 44 19E
Annam Range *mts.* Laos/Vietnam **60** B4 19 00N 104 00E
Annan Scotland **33** F2 54 59N 3 16W
Annan *r.* Scotland **33** F3 55 05N 3 20W
Annapurna *mt.* Nepal **58** D3 28 34N 83 50E
Annecy France **48** G2 45 54N 6 07E
Anshan China **59** D4 41 05N 122 58E
Anstruther Scotland **33** G4 56 14N 2 42W
Antalya Turkey **43** N3 36 53N 30 42E
Antananarivo Madagascar **67** E3 18 52S 47 30E
Antarctic Peninsula Antarctica **86** 68 00S 65 00W
Antibes France **48** G1 43 35N 7 07E
Antigua *i.* Antigua & Barbuda **84** C9 17 09N 61 49W
ANTIGUA AND BARBUDA 79 N2
Antofagasta Chile **84** B5 23 40S 70 23W
Antrim Northern Ireland **32** C2 54 43N 6 13W
Antrim *district* Northern Ireland **32** C2 54 45N 6 25W
Antrim Mountains Northern Ireland **32** C3/D2 55 00N 6 10W
Antwerp Belgium **49** C4 51 13N 4 25E
Anxi China **59** A5 40 32N 95 57E
Aomori Japan **59** N3 40 50N 140 43E
Aosta Italy **51** A6 45 43N 7 19E
Aparri The Philippines **60** D4 18 22N 121 40E
Apeldoorn Netherlands **49** C5 52 13N 5 57E
Appalachians *mts.* USA **79** L4 37 00N 82 00W
Appennines *mts.* Italy **51** C6/F4 44 30N 10 00E
Appleby-in-Westmorland England **34** C3 53 36N 2 29W
Aqaba Jordan **61** D3 29 32N 35 00E
Arabian Sea Indian Ocean **7** 17 00N 60 00E
Aracaju Brazil **84** F6 10 54S 37 07W
Arafura Sea Australia **72** D5 9 00S 133 00E
Araguaia *r.* Brazil **84** D6 12 30S 51 00W
Arak Iran **61** E4 34 05N 49 42E
Araks *r.* Asia **61** E4 39 30N 48 00E
Aral Sea Asia **56** G2 45 00N 60 00E
Aran Fawddy *mt.* Wales **36** D2 52 47N 3 41W
Ararat, Mount Turkey **43** Q3 39 44N 44 15E
Arbil Iraq **61** E4 36 12N 44 01E
Arbroath Scotland **31** G1 56 34N 2 35W
Arctic Ocean 86
Ardabil Iran **61** E4 38 15N 48 18E

Ardbeg Scotland **32** C3 55 39N 6 05W
Ardennes *mts.* Belgium **49** C4 50 10N 5 45E
Ardminish Scotland **32** D3 55 42N 5 44W
Ardres France **39** F1 50 51N 1 59E
Ards *district* Northern Ireland **32** D2 54 30N 5 30W
Ards Peninsula Northern Ireland **32** D2 54 25N 5 30W
Arequipa Peru **84** B6 16 25S 71 32W
Arezzo Italy **51** C5 43 28N 11 53E
ARGENTINA 85 C4
Argyll and Bute *u.a.* Scotland **32/33** D4/E4 56 10N 5 00W
Århus Denmark **42** J7 56 15N 10 10E
Arica Chile **84** B6 18 30S 70 20W
Arisaig Scotland **30** D1 56 51N 5 51W
Arisaig, Sound of Scotland **30** D1 56 50N 5 50W
Arizona *state* USA **79** H4 34 00N 112 00W
Arkansas *r.* USA **79** K4 34 00N 93 00W
Arkansas *state* USA **79** K4 34 00N 93 00W
Arkhangel'sk Russia **43** T6 64 32N 40 40E
Arklow Republic of Ireland **36** A2 52 48N 6 09W
Arles France **48** F1 43 41N 4 38E
Armagh Northern Ireland **32** C2 54 21N 6 39W
Armagh *district* Northern Ireland **32** C2 54 20N 6 40W
ARMENIA 43 Q4
Armentières France **39** G1 50 41N 2 53E
Arnhem Netherlands **49** C4 52 00N 5 53E
Arno *r.* Italy **51** C5 43 45N 11 00E
Arnold England **37** F2 53 00N 1 09W
Arran *i.* Scotland **32** D3 55 35N 5 15W
Arras France **48** E5 50 17N 2 46E
Artá Balearic Islands **50** G3 39 42N 3 20E
Aruba *i.* Netherlands **79** N2 12 20N 70 00W
Aru Islands Indonesia **60** E2 6 00S 134 30E
Arun *r.* England **38** D2 51 05N 0 30W
Arundel England **38** D1 50 51N 0 34W
Asahikawa Japan **59** N3 43 46N 142 23E
Asahi, Mount Japan **59** N3 43 42N 142 54E
Asansol India **58** D2 23 40N 86 59E
Ascension Island Atlantic Ocean **5** 7 57S 14 22W
Ashbourne England **37** F3 53 01N 1 43W
Ashburton New Zealand **73** C2 43 54S 171 45E
Ashford England **39** E2 51 09N 0 53E
Ashgabat Turkmenistan **56** F1 37 58N 58 24E
Ashington England **33** H3 55 11N 1 34W
Asmara Eritrea **61** D3 15 20N 38 58E
Assisi Italy **51** D5 43 04N 12 37E
Astana Kazakhstan **56** H3 51 10N 71 28E
Astrakhan Russia **56** E2 46 22N 48 04E
Asunción Paraguay **85** D5 25 15S 57 40W
Aswan Egypt **61** D3 24 05N 32 56E
Asyut Egypt **61** D3 27 14N 31 07E
Atacama Desert Chile **84/85** B5 22 30S 70 00W
Atbara Sudan **61** D2 17 42N 34 00E
Atbara *r.* Sudan **61** D2 17 00N 35 00E
Athabasca *r.* Canada **78** H6 57 30N 112 00W
Athens Greece **42** L3 38 00N 23 44E
Atlanta USA **79** L4 33 45N 84 23W
Atlantic Ocean 79 N4
Atlas Mountains Morocco **66** B6 32 00N 2 00W
Atrek *r.* Iran/Turkmenistan **61** F4 38 00N 57 00E
Atyrau Kazakhstan **56** F2 47 08N 51 59E
Auchterarder Scotland **31** F1 56 18N 3 43W
Auchtermuchty Scotland **33** F4 56 17N 3 14W
Auckland New Zealand **73** C3 36 51S 174 46E
Auderville France **41** F2 49 43N 1 56W
Augsburg Germany **49** F3 48 21N 10 54E
Austin USA **79** K4 30 18N 97 47W
AUSTRALIA 72
AUSTRIA 42 J5
Auxerre France **48** E3 47 48N 3 35E
Aveiro Portugal **50** A4 40 38N 8 40W
Aviemore Scotland **31** F2 57 12N 3 50W
Avignon France **48** F1 43 56N 4 48E
Avon *r.* Warwickshire/Worcestershire England **37** F2 52 15N 1 55W
Avon *r.* Wiltshire/Bath and North East Somerset/ North Somerset England **38** B2 51 20N 2 05W
Avon *r.* Wiltshire/Hampshire/Dorset England **38** C1 50 55N 1 50W
Axe *r.* England **41** E3 50 50N 2 50W
Axminster England **41** E3 50 47N 3 00W
Ayaguz Kazakhstan **56** J2 47 59N 80 27E
Ayers Rock *mt.* Australia **72** D3 25 20S 131 01E
Aylesbury England **38** D2 51 50N 0 49W
Ayr Scotland **33** E3 55 28N 4 38W
Ayr *r.* Scotland **33** E3 55 30N 4 20W
Ayre, Point of *c.* Isle of Man British Isles **33** E2 54 25N 4 22W
AZERBAIJAN 56 E2
Azores *is.* Atlantic Ocean **4** 38 30N 28 00W
Azov, Sea of Asia **43** P5 46 00N 36 00E

B

Bacolod The Philippines **60** D4 10 38N 122 58E
Badajoz Spain **50** B3 38 53N 6 58W
Baden-Baden Germany **49** E3 48 45N 8 15E
Baffin Bay Canada **78** N8 72 00N 64 00W
Baffin Island Canada **78** M8 70 00N 75 00W
Baghdad Iraq **61** E4 33 20N 44 26E
BAHAMAS, THE 79 M3
Bahia Blanca Argentina **85** C4 38 45S 62 15W
BAHRAIN 61 F3
Baie de la Seine *b.* France **41** G2 49 40N 0 30W
Bakewell England **37** F3 53 13N 1 40W
Baku Azerbaijan **56** E2 40 22N 49 53E
Bala Wales **36** D2 52 54N 3 35W
Bala Lake Wales **36** D2 52 53N 3 38W
Balearic Islands Spain **50** F3/H4 40 00N 2 00E
Bali *i.* Indonesia **60** C2 8 30S 115 00E
Balikesir Turkey **42** M3 39 37N 27 51E
Balikpapan Indonesia **60** C2 1 15S 116 50E
Balkhash, Lake Kazakhstan **56** H2 46 00N 75 00E
Ballantrae Scotland **32** D3 55 06N 5 00W
Ballarat Australia **72** E2 37 33S 143 52E
Ballater Scotland **31** F2 57 03N 3 03W
Ballybofey Republic of Ireland **32** B2 54 48N 7 47W
Ballycastle Northern Ireland **32** C3 55 12N 6 15W
Ballymena Northern Ireland **32** C2 54 52N 6 17W

Ballymena *district* Northern Ireland **32** C2 54 55N 6 20W
Ballymoney Northern Ireland **32** C3 55 04N 6 31W
Ballymoney *district* Northern Ireland **32** C3 55 00N 6 30W
Ballyshannon Republic of Ireland **32** A2 54 30N 8 11W
Baltic Sea Europe **42** K7 55 15N 17 00E
Baltimore USA **79** M4 39 18N 76 38W
Bamako Mali **66** B4 12 40N 7 59W
Bamberg Germany **49** F3 49 54N 10 54E
Bamburgh England **33** H3 55 36N 1 42W
Banbridge Northern Ireland **32** C2 54 21N 6 16W
Banbridge *district* Northern Ireland **32** C2 54 20N 6 20W
Banbury England **38** C3 52 04N 1 20W
Banchory Scotland **31** G2 57 30N 2 30W
Banda Aceh Indonesia **60** A3 5 30N 95 20E
Bandar-e 'Abbas Iran **61** F3 27 12N 56 15E
Bandar Seri Begawan Brunei **60** C3 4 53N 114 57E
Banda Sea Indonesia **60** D2 5 50S 126 00E
Bandung Indonesia **60** B2 6 57S 107 34E
Banff Scotland **31** G2 57 40N 2 33W
Bangalore India **58** C2 12 58N 77 35E
Bangka *i.* Indonesia **60** B2 2 30S 107 00E
Bangkok Thailand **60** B4 13 44N 100 30E
BANGLADESH 58 D3/E3
Bangor Northern Ireland **32** D2 54 40N 5 40W
Bangor Wales **36** C3 53 13N 4 08W
Bangui Central African Republic **67** B5 4 23N 18 37E
Bangweulu, Lake Zambia **67** C3 11 15S 29 45E
Banjarmasin Indonesia **60** C2 3 22S 114 36E
Banjul The Gambia **66** A4 13 28N 16 39W
Banks Island Canada **78** G8 73 15N 121 30W
Bann *r.* Northern Ireland **32** C2 54 55N 6 30W
Baotou China **59** B4 40 38N 109 59E
BARBADOS 79 P2
Barbuda *i.* Antigua & Barbuda **84** C9 17 41N 61 48W
Barcelona Spain **50** G4 41 25N 2 10E
Bareilly India **58** C3 28 20N 79 24E
Barents Sea Arctic Ocean **86** 75 00N 40 00E
Barfleur France **41** G2 49 40N 1 16W
Bari Italy **51** F4 41 07N 16 52E
Barito *r.* Indonesia **60** C2 2 00S 115 00E
Barking England **39** E2 51 33N 0 06E
Barletta Italy **51** F4 41 20N 16 17E
Barmouth Wales **36** C2 52 43N 4 03W
Barnard Castle England **34** D3 54 33N 1 55W
Barnaul Russia **56** J3 53 21N 83 45E
Barnet England **38** D2 51 39N 0 12W
Barnoldswick England **34** C2 53 56N 2 16W
Barnsley England **35** D2 53 34N 1 28W
Barnstaple England **40** C4 51 05N 4 04W
Barquisimeto Venezuela **84** C9 10 03N 69 18W
Barra *i.* Scotland **30** B1/2 57 00N 7 25W
Barranquilla Colombia **84** B9 11 10N 74 50W
Barreiras Brazil **84** E6 12 09S 44 58W
Barrow-in-Furness England **34** B3 54 07N 3 14W
Barry Wales **36** D1 51 24N 3 18W
Barton-upon-Humber England **35** E2 53 41N 0 27W
Basel Switzerland **49** D2 47 33N 7 36E
Basildon England **39** E2 51 34N 0 25W
Basingstoke England **38** C2 51 16N 1 05W
Basra Iraq **61** E4 30 30N 47 50E
Bassein Myanmar **58** E2 16 46N 94 45E
Bass Strait Australia **72** E2 39 51S 145 53E
Bastia France **51** B5 42 14N 9 26E
Bata Equatorial Guinea **66** C3 1 51N 9 49E
Batang China **59** A3 30 02N 99 01E
Batangas The Philippines **60** D4 13 46N 121 01E
Bath England **38** B2 51 23N 2 22W
Bath and North East Somerset *u.a.* England **38** B2 51 25N 2 30W
Bathgate Scotland **33** F3 55 55N 3 39W
Baton Rouge USA **79** K4 30 30N 91 10W
Batumi Georgia **56** E2 41 37N 41 36E
Bavaria *admin.* Germany **49** F3/G3 49 00N 12 00E
Bayeux France **48** C4 49 16N 0 42W
Baykal, Lake Russia **57** L4 54 00N 109 00E
Bayonne France **48** C1 43 30N 1 28W
Bayreuth Germany **49** F3 49 27N 11 35E
Beachy Head *c.* England **39** E1 50 44N 0 16E
Bearsden Scotland **33** E3 55 56N 4 20W
Beaufort Sea Arctic Ocean **78** E7 72 00N 140 00W
Beauly *r.* Scotland **31** E2 57 30N 4 30W
Beauvais France **48** E4 49 26N 2 05E
Beccles England **39** F2 52 28N 1 34E
Bechar Algeria **66** B6 31 35N 2 17W
Bedford England **38** D3 52 08N 0 29W
Bedfordshire *co.* England **38** D3 52 08N 0 29W
Bedworth England **37** F2 52 29N 1 28W
Beersheba Israel **61** D4 31 15N 34 47E
Beida Libya **66** E6 32 00N 21 30E
Beijing China **59** C3 39 55N 116 26E
Beinn Dearg *mt.* Scotland **31** E2 57 47N 4 56W
Beira Mozambique **67** D3 19 49S 34 52E
Beirut Lebanon **61** D4 33 52N 35 30E
Beja Portugal **50** B3 38 01N 7 52W
Bejaïa Algeria **42** H3 36 49N 5 03E
BELARUS 56 C3
Belém Brazil **84** E7 1 27S 48 29W
Belfast Northern Ireland **32** D2 54 35N 5 55W
Belfast *district* Northern Ireland **32** C2/D2 54 30N 5 58W
Belfast Lough *est.* Northern Ireland **32** D2 54 40N 5 45W
Belfort France **48** G3 47 38N 6 52E
Belgaum India **58** C2 15 54N 74 36E
BELGIUM 49
Belgrade Serbia and Montenegro **42** L4 44 50N 20 30E
Belitung *i.* Indonesia **60** B2 2 30S 107 30E
BELIZE 79 L2
Belize Belize **79** L2 17 29N 88 10W
Belle Isle *i.* France **48** B3 47 20N 3 10W
Bellingshausen Sea Southern Ocean **86** 71 00S 85 00W
Belmopan Belize **79** L2 17 13N 88 48W
Belo Horizonte Brazil **84** E6 19 54S 43 54W
Ben Alder *mt.* Scotland **31** E1 56 49N 4 28W

Benavente Spain **50** C5 42 00N 5 40W
Benbecula *i.* Scotland **30** B2 57 25N 7 20W
Ben Cruachan *mt.* Scotland **30** D1 56 26N 5 09W
Bengal, Bay of Indian Ocean **58** D2 17 00N 88 00E
Benghazi Libya **66** E6 32 07N 20 04E
Benguela Angola **67** B3 12 34S 13 24E
Ben Hope *mt.* Scotland **31** E3 58 24N 4 37W
BENIN 66 C4
Ben Klibreck *mt.* Scotland **31** E3 58 14N 4 22W
Ben Lawers *mt.* Scotland **31** E1 56 33N 4 15W
Ben Lomond *mt.* Scotland **33** E4 56 12N 4 38W
Ben More *mt.* Argyll and Bute Scotland **30** C1 56 25N 6 02W
Ben More *mt.* Stirling Scotland **31** E1 56 25N 6 02W
Ben More Assynt *mt.* Scotland **31** E3 58 07N 4 52W
Ben Nevis *mt.* Scotland **31** D1 56 48N 5 00W
Benue *r.* Nigeria/Cameroon **66** C3/D3 9 00N 10 00E
Ben Wyvis *mt.* Scotland **31** E2 57 40N 4 35W
Berbera Somalia **66** E6 10 28N 45 02E
Bereziniki Russia **43** T7 59 26N 56 49E
Bérgamo Italy **51** B6 45 42N 9 40E
Bergen Norway **42** H8 60 23N 5 20E
Bergerac France **48** D2 44 50N 0 29E
Bering Sea Pacific Ocean **57** S3 58 00N 170 00W
Bering Strait Russia/USA **78** C7 69 00N 169 00W
Berlin Germany **49** G5 52 32N 13 25E
Bermuda *i.* Atlantic Ocean **79** N4 32 50N 64 20W
Bern Switzerland **49** D2 46 57N 7 26E
Berneray *i.* Scotland **30** B2 57 45N 7 10W
Bernina *mt.* Europe **49** E2 46 23N 9 54E
Berwick-upon-Tweed England **33** G3 55 46N 2 00W
Besançon France **48** F3 47 14N 6 02E
Bethesda Wales **36** C3 53 11N 4 03W
Béthune France **48** E5 50 32N 2 38E
Beverley England **35** E2 53 51N 0 26W
Bewdley England **37** E2 52 22N 2 19W
Bexley England **39** E2 51 27N 0 09E
BHUTAN 58 D3/E3
Białystok Poland **42** L6 53 09N 23 10E
Biarritz France **48** C1 43 29N 1 33W
Bicester England **38** C2 51 54N 1 09W
Bideford England **40** C4 51 01N 4 13W
Bideford Bay England **40** C4 51 05N 4 25W
Bielefeld Germany **49** E5 52 02N 8 32E
Bigbury Bay England **40** D3 50 17N 4 00W
Biggar Scotland **33** F3 55 38N 3 32W
Biggleswade England **38** D3 52 05N 0 17W
Bikaner India **58** C3 28 01N 73 22E
Bilbao Spain **50** D5 43 15N 2 56W
Bintan *i.* Indonesia **60** B3 0 50N 104 30E
Birdum Australia **72** D4 15 39S 133 13E
Birkenhead England **34** B2 53 24N 3 02W
Birmingham England **37** F2 52 30N 1 50W
Birmingham USA **79** L4 33 30N 86 55W
Biscay, Bay of Atlantic Ocean **48** B2 45 30N 2 50W
Bishkek Kyrgyzstan **56** H2 42 53N 74 46E
Bishop Auckland England **34** D3 54 40N 1 40W
Bishop's Stortford England **39** E2 51 53N 0 09E
Biskra Algeria **42** H2 34 50N 5 41E
Bissau Guinea-Bissau **66** A4 11 52N 15 39W
Biwa, Lake Japan **59** M2 35 20N 135 20E
Bizerte Tunisia **51** B2 37 18N 9 52E
Blackburn England **34** C2 53 45N 2 29W
Blackburn with Darwen *u.a.* England **34** C2 53 43N 2 25W
Black Forest *mts.* Germany **49** E3 47 00N 8 00E
Black Mountains Wales **36** D1 51 55N 3 10W
Blackpool England **34** B2 53 50N 3 03W
Blackpool *u.a.* England **34** B2 53 51N 3 10W
Black Sea Europe **43** P4 43 00N 35 00E
Black Volta *r.* Africa **66** B3/4 10 00N 3 00W
Blackwater *r.* Northern Ireland **32** C2 54 30N 6 45W
Blackwater Reservoir Scotland **30** E1 56 41N 4 46W
Blaenau Ffestiniog Wales **36** D2 52 59N 3 56W
Blaenau Gwent *u.a.* Wales **36** D1 51 45N 3 10W
Blagoveshchensk Russia **57** N3 50 19N 127 30E
Blairgowrie Scotland **31** F1 56 36N 3 21W
Blandford Forum England **41** E3 50 52N 2 11W
Blantyre Malawi **67** D3 15 46S 35 00E
Blenheim New Zealand **73** C2 41 31S 173 57E
Blida Algeria **42** G3 36 30N 2 50E
Blois France **48** D3 47 36N 1 20E
Bloemfontein Republic of South Africa **67** C2 29 07S 26 14E
Blue Nile *r.* Ethiopia/Sudan **61** D2 12 00N 33 00E
Blyth England **33** H3 55 07N 1 30W
Blyth *r.* England **33** H3 55 10N 1 35W
Boa Vista Brazil **84** C8 3 23S 55 30W
Bobo Dioulasso Burkina **66** B4 11 11N 4 18W
Bochum Germany **49** D4 51 29N 7 13E
Bodele Depression Chad **66** D4 17 00N 17 50E
Bodmin England **40** C3 50 29N 4 43W
Bodmin Moor England **40** C3 50 35N 4 40W
Bodø Norway **42** J9 67 18N 14 26E
Bognor Regis England **38** D1 50 47N 0 41W
Bogor Indonesia **60** B2 6 34S 106 45E
Bogotá Colombia **84** B8 4 38N 74 05W
Boise USA **79** H5 43 38N 116 12W
BOLIVIA 84 C6
Bologna Italy **51** C6 44 30N 11 20E
Bolton England **34** C2 53 35N 2 26W
Bolzano Italy **51** C7 46 30N 11 22E
Bonaire *i.* Lesser Antilles **79** N2 12 15N 68 27W
Bonar Bridge *tn.* Scotland **31** E2 57 53N 4 21W
Bo'ness Scotland **33** F4 56 01N 3 37W
Bonifacio Corsica **51** B4 41 23N 9 10E
Bonn Germany **49** D4 50 44N 7 06E
Bootle England **34** B2 53 28N 3 00W
Bordeaux France **48** C2 44 50N 0 34W
Borneo *i.* Indonesia/Malaysia **60** C3 1 00N 113 00E
Bornholm *i.* Denmark **49** H6 55 02N 15 00E
Boscastle England **40** C3 50 41N 4 42W
BOSNIA-HERZEGOVINA 42 K4
Boston England **35** E1 52 29N 0 01W
Boston USA **79** M5 42 20N 71 05W
Bothnia, Gulf of Finland/Sweden **42** L8 61 00N 19 10E
BOTSWANA 67 C2

Bouaké Côte d'Ivoire **66** B3 7 42N 5 00W
Bougainville *i.* Papua New Guinea **72** F5 6 15S 155 00E
Boulogne-sur-Mer France **48** D5 50 43N 1 37E
Bourges France **48** E3 47 05N 2 23E
Bourke Australia **72** E2 30 05S 145 58E
Bournemouth England **41** F3 50 43N 1 54W
Bournemouth *u.a.* England **41** F3 50 46N 1 52W
Bou Saâda Algeria **42** G3 35 10N 4 09E
Bovey Tracey England **40** D3 50 36N 3 40W
Boyoma Falls Congo Dem.Rep. **67** C5 0 18N 25 30E
Bracknell England **38** D2 51 26N 0 46W
Bracknell Forest *u.a.* England **38** D2 51 22N 0 43W
Bradford England **34** D2 53 48N 1 45W
Braemar Scotland **31** F2 57 01N 3 23W
Brahmaputra *r.* India/Bangladesh **58** E3 27 00N 94 00E
Brăila Romania **42** M5 45 17N 27 58E
Braintree England **39** E2 51 53N 0 32E
Brampton England **34** C3 54 57N 2 43W
Brandenburg Germany **49** G5 52 25N 12 34E
Brandenburgh Scotland **49** F3 52 43N 3 18W
Brasília Brazil **84** E6 15 45S 47 57W
Braşov Romania **42** M5 45 39N 25 35E
Bratislava Slovakia **42** K5 48 10N 17 10E
Bratsk Russia **57** L3 56 20N 101 50E
Braunton England **40** C4 51 07N 4 10W
Bray Republic of Ireland **36** A3 53 12N 6 06W
BRAZIL 84 C7/E7
Brazilian Plateau Brazil **84** E6 17 00S 44 00W
Brazzaville Congo **66** C2 4 14S 15 14E
Brechin Scotland **31** G2 56 44N 2 40W
Brecon Wales **36** D1 51 57N 3 24W
Brecon Beacons *mts.* Wales **36** D1 51 53N 3 30W
Breda Netherlands **49** C4 51 35N 4 46E
Bremen Germany **49** E5 53 05N 8 48E
Bremerhaven Germany **49** E5 53 33N 8 35E
Brenner Pass Austria/Italy **49** F2 47 02N 11 32E
Brent England **38** D2 51 34N 0 17W
Brentwood England **39** E2 51 38N 0 18E
Bréscia Italy **51** C6 45 33N 10 13E
Bressay *i.* Scotland **29** D3 60 08N 1 05W
Brest Belarus **42** L6 52 08N 23 40E
Brest France **48** A4 48 23N 4 30W
Bridgend Wales **36** D1 51 31N 3 35W
Bridgend *u.a.* Wales **36** D1 51 32N 3 35W
Bridgnorth England **37** E2 52 33N 2 25W
Bridgwater England **41** E4 51 08N 3 00W
Bridgwater Bay England **41** D4 51 15N 3 20W
Bridlington England **35** E3 54 05N 0 12W
Bridport England **41** E3 50 44N 2 46W
Brigg England **35** E2 53 34N 0 30W
Brighouse England **34** D2 53 42N 1 47W
Brighton England **38** D1 50 50N 0 10W
Brighton & Hove *u.a.* England **38** D1 50 52N 0 13W
Brindisi Italy **51** F4 40 37N 17 57E
Brisbane Australia **72** F3 27 28S 153 03E
Bristol England **38** B2 51 27N 2 35W
Bristol England *u.a.* **38** B2 51 30N 2 30W
Bristol Channel United Kingdom **36** C1/D1 51 20N 3 50W
British Columbia *province* Canada **78** G6 55 00N 128 00W
Brittany *geog. reg.* France **48** B4 48 00N 3 00W
Brive France **48** D2 45 09N 1 32E
Brixham England **40** D3 50 23N 3 30W
Brno Czech Republic **42** K5 49 13N 16 40E
Broad Bay Scotland **30** C3 58 15N 6 10W
Broadford Scotland **30** D2 57 14N 5 54W
Broad Law *mt.* Scotland **33** F3 55 30N 3 22W
Broadway England **37** F2 52 02N 1 50W
Brodick Scotland **32** D3 55 35N 5 09W
Broken Hill *tn.* Australia **72** E2 31 57S 141 28E
Bromley England **39** E2 51 31N 0 01E
Bromsgrove England **37** E2 52 20N 2 03W
Bromyard England **37** E2 52 11N 2 30W
Brooks Range *mts.* U.S.A. **78** D7/E7 67 55N 155 00W
Broome Australia **72** C4 17 57S 122 13E
Brora Scotland **31** F3 58 01N 3 51W
Brough England **34** C3 54 32N 2 19W
Brough Head Scotland **29** B2 59 09N 3 19W
Brown Clee Hill England **37** E2 52 27N 2 35W
Brown Willy *sum.* England **40** C3 50 35N 4 36W
Brugge Belgium **49** B4 51 13N 3 14E
BRUNEI 60 C3
Brunswick Germany **49** F5 52 15N 10 30E
Brussels Belgium **49** C4 50 50N 4 21E
Bryansk Russia **56** D3 53 15N 34 09E
Bryher *i.* England **40** A2 49 57N 6 21W
Bucaramanga Colombia **84** B8 7 08N 73 10W
Buchan Ness *c.* Scotland **31** H2 57 28N 1 47W
Bucharest Romania **42** M4 44 25N 26 07E
Buckfastleigh England **40** D3 50 29N 3 46W
Buckhaven Scotland **33** F4 56 11N 3 03W
Buckie Scotland **31** G2 57 40N 2 58W
Buckingham England **38** D2 52 00N 1 00W
Buckinghamshire *co.* England **38** D2 51 50N 0 50W
Bude England **40** C3 50 50N 4 33W
Bude Bay England **40** C3 50 50N 4 40W
Buenaventura Colombia **84** B8 3 54N 77 02W
Buenos Aires Argentina **85** D4 34 40S 58 30W
Buffalo USA **79** M5 42 52N 78 55W
Builth Wells Wales **36** D2 52 09N 3 24W
Bujumbura Burundi **67** C4 3 22S 29 19E
Bukavu Congo Dem. Rep. **67** C4 2 30S 28 50E
Bukhara Uzbekistan **56** G1 39 47N 64 26E
Bulawayo Zimbabwe **67** C2 20 10S 28 43E
BULGARIA 42 M4
Bunbury Australia **72** B2 33 20S 115 39E
Buncrana Republic of Ireland **32** B3 55 08N 7 27W
Bundaberg Australia **72** F3 24 53S 152 21E
Bungay England **39** F3 52 28N 1 26E
Buraida Saudi Arabia **61** E3 26 20N 43 59E
Bure *r.* England **39** F3 52 47N 1 20E
Burgas Bulgaria **42** M4 42 30N 27 29E
Burghead Scotland **31** F2 57 42N 3 30W
Burgos Spain **50** D5 42 21N 3 41W
Burgundy *geog. reg.* France **48** F3 47 00N 5 00E
BURKINA 66 B4

BURMA *see* MYANMAR
Burnie Australia **72** E1 41 03S 145 55E
Burnley England **34** C2 53 48N 2 14W
Burry Port Wales **36** C1 51 42N 4 15W
Bursa Turkey **42** M4 40 12N 29 04E
Burton upon Trent England **37** F2 52 48N 1 36W
Buru *i.* Indonesia **60** D2 3 30S 126 30E
BURUNDI **67** C4
Bury England **34** C2 53 36N 2 17W
Bury St. Edmunds England **39** E3 52 15N 0 43E
Bush *r.* Northern Ireland **32** C3 55 10N 6 30W
Bute *i.* Scotland **32** D3 55 50N 5 05W
Bute, Sound of Scotland **32** D3 55 45N 5 10W
Buton *i.* Indonesia **60** D2 5 00S 122 45E
Butt of Lewis *c.* Scotland **30** C3 58 30N 6 20W
Butuan The Philippines **60** D3 8 56N 125 31E
Buxton England **37** F3 53 15N 1 55W
Bydgoszcz Poland **42** K6 53 16N 18 00E

C

Cabanatuan The Philippines **60** D4 15 30N 120 58E
Cabinda *province* Angola **66** D2 5 00S 12 00E
Cabora Bassa, Lake Mozambique **67** D3 15 40S 31 40E
Cabrera *i.* Balearic Islands **50** G3 39 00N 2 59E
Cader Idris *mt.* Wales **36** D2 52 42N 3 54W
Cádiz Spain **50** B2 36 32N 6 18W
Caen France **48** C4 49 11N 0 22W
Caernarfon Wales **36** C3 53 08N 4 16W
Caernarfon Bay Wales **36** C3 53 05N 4 30W
Caerphilly Wales **36** D1 51 35N 3 14W
Caerphilly *u.a.* Wales **36** D1 51 37N 3 15W
Cagayan de Oro The Philippines **60** D3 8 29N 124 40E
Cágliari Italy **51** B3 39 13N 9 08E
Cairn Gorm *mt.* Scotland **31** F2 57 07N 3 40W
Cairngorms *mts.* Scotland **31** F2 57 10N 3 30W
Cairns Australia **72** E4 16 54S 145 45E
Cairo Egypt **61** D4 30 03N 31 15E
Calais France **48** D5 50 57N 1 52E
Cala Ratjada Balearic Islands **50** G3 39 43N 3 27E
Calatayud Spain **50** E4 41 21N 1 39W
Calbayog The Philippines **60** D4 12 04N 124 38E
Calgary Canada **78** F6 51 05N 114 05W
Cali Colombia **84** B8 3 24N 76 30W
Calicut India **58** C2 11 15N 75 45E
California *state* USA **79** H4 35 00N 119 00W
California, Gulf of Mexico **79** H3 27 00N 111 00W
Callander Scotland **33** E4 56 15N 4 13W
Callao Peru **84** B6 12 05S 77 08W
Calligarry Scotland **30** D2 57 02N 5 58E
Calne England **38** B2 51 27N 2 00W
Cam *r.* England **39** E3 52 15N 0 15W
Camagüey Cuba **79** M3 21 25N 77 55W
Camberley England **38** D2 51 21N 0 45W
CAMBODIA **60** B4
Camborne England **40** B3 50 12N 5 19W
Cambrai France **48** E5 50 10N 3 14E
Cambrian Mountains Wales **36** D2 52 15N 3 45W
Cambridge England **39** E3 52 12N 0 07E
Cambridgeshire *co.* England **38/39** D3/E3 52 30N 0 00
Camel *r.* England **40** C3 50 30N 4 45W
CAMEROON **66** D3
Cameroun, Mount Cameroon **66** C3 4 13N 9 10E
Campbeltown Scotland **32** D3 55 26N 5 36W
Campina Grande Brazil **84** F7 7 15S 35 50W
Campinas Brazil **84** E5 22 54S 47 06W
Campo Grande Brazil **84** D5 20 24S 54 35W
Campos Brazil **84** E5 21 46S 41 21W
Campsie Fells *hills* Scotland **33** E4 56 00N 4 15W
CANADA **78**
Canary Islands Spain **66** A5 28 30N 15 10W
Canaveral, Cape USA **79** L3 28 28N 80 28W
Canberra Australia **72** E2 35 17S 149 09E
Canisp *mt.* Scotland **31** D3 58 07N 5 03W
Canna *i.* Scotland **30** C2 57 05N 6 35W
Cannes France **48** G1 43 33N 7 00E
Cannock England **37** E2 52 42N 2 02W
Cantabrian Mountains Spain **50** B5/C5 43 00N 5 30W
Canterbury England **39** F2 51 17N 1 05E
Canterbury Bight *b.* New Zealand **73** C2 44 00S 172 00E
Canterbury Plains New Zealand **73** C2 43 45S 171 56E
Can Tho Vietnam **60** B4 10 03N 105 46E
Cap de la Hague *c.* France **41** F2 49 44N 1 56W
Cape Town Republic of South Africa **67** B1 33 56S 18 28E
CAPE VERDE **4**
Cap Gris-Nez *c.* France **39** F1 50 52N 1 35E
Caracas Venezuela **84** C9 10 35N 66 56W
Carcassonne France **48** E1 43 13N 2 21E
Cardiff Wales **36** D1 51 30N 3 13W
Cardiff *u.a.* Wales **36** D1 51 32N 3 14W
Cardigan Wales **36** C2 52 06N 4 40W
Cardigan Bay Wales **36** C2 52 30N 4 30W
Carentan France **41** F2 49 19N 1 14W
Caribbean Sea Central America **79** M2 15 00N 75 00W
Carlingford Lough *est.* Northern Ireland/ Republic of Ireland **32** C2 54 05N 6 10W
Carlisle England **34** C3 54 54N 2 55W
Carmarthen Wales **36** C1 51 51N 4 20W
Carmarthen Bay Wales **36** C1 51 40N 4 30W
Carmarthenshire *u.a.* Wales **36** C1/D1 51 55N 4 00W
Carmel Head Wales **36** C3 53 24N 4 34W
Carn Eige *mt.* Scotland **31** D2 57 22N 5 07W
Carnforth England **34** C3 54 08N 2 46W
Carnoustie Scotland **31** G1 56 30N 2 44W
Carpathians *mts.* Europe **42** L5/M5 49 00N 22 00E
Carpentaria, Gulf of Australia **72** D4 14 39S 139 11E
Carrickfergus Northern Ireland **32** D2 54 43N 5 49W
Carrickfergus *district* Northern Ireland **32** C2 54 40N 5 50W
Cartagena Colombia **84** B9 10 24N 75 33W
Cartagena Spain **50** E2 37 36N 0 59W
Carteret France **41** F2 49 22N 1 48W
Casablanca Morocco **66** B6 33 39N 7 35W
Caspian Sea Asia **56** F2 41 00N 50 00E
Cassel France **39** G1 51 40N 2 28E

Cassino Italy **51** D4 41 29N 13 50E
Castlebay Scotland **30** B1 56 57N 7 28W
Castleblaney Republic of Ireland **32** C2 54 07N 6 44W
Castle Douglas Scotland **33** F2 54 57N 3 56W
Castleford England **35** D2 43 44N 1 21W
Castlereagh *district* Northern Ireland **32** D2 54 35N 5 50W
Castletown Isle of Man British Isles **33** E2 54 04N 4 38W
Castrovillari Italy **51** F3 39 48N 16 12E
Catánia Italy **51** E2 37 31N 15 06E
Catanzaro Italy **51** F3 38 54N 16 36E
Cat Island The Bahamas **79** M3 24 30N 75 30W
Cauca *r.* Colombia **84** B8 7 40N 75 00W
Caucasus *mts.* Asia **56** F2 43 00N 43 00E
Cavan Republic of Ireland **32** B1 53 58N 7 21W
Caxias do Sul Brazil **85** D5 29 14S 51 10W
Cayenne French Guiana **84** D8 4 55N 52 18W
Cayman Islands Caribbean Sea **79** L2 19 20N 81 15W
Cebollera *mt.* Spain **50** D4 42 00N 2 30W
Cebu The Philippines **60** D4 10 17N 123 56E
Celebes Sea Indonesia **60** D3 3 00N 122 00E
Cemaes Head Wales **36** C2 53 10N 4 50W
CENTRAL AFRICAN REPUBLIC **67** B5/C5
Central Siberian Plateau Russia **57** L4 65 00N 110 00E
Ceredigion *u.a.* Wales **36** C2/D2 52 20N 4 00W
Ceuta *territory* Spain **50** C1 35 53N 5 19W
CHAD **66** D4
Chad, Lake West Africa **66** D4 13 50N 14 00E
Chagos Archipelago *is.* Indian Ocean **5** 6 00S 72 00E
Châlons-sur-Marne France **48** F4 48 58N 4 22E
Chalon-sur-Saône France **48** F3 46 47N 4 51E
Chambéry France **48** F2 45 34N 5 55E
Changchun China **59** D4 43 50N 125 20E
Chang Jiang (*Yangtze*) *r.* China **59** C3 30 00N 116 00E
Changsha China **59** C2 28 10N 113 00E
Channel Islands British Isles **41** E2 49 30N 2 30W
Channel Tunnel United Kingdom/France **39** F1/2 5 00N 1 30E
Chard England **41** E3 50 53N 2 58W
Chardzhev Turkmenistan **56** G1 39 09N 63 34E
Chari *r.* Chad/Sudan **66** D4 11 00N 16 00E
Charleroi Belgium **49** C4 50 25N 4 27E
Charlestown of Aberlour Scotland **31** F2 57 28N 3 14W
Charleville Australia **72** E3 26 24S 146 14E
Charleville France **48** F4 49 46N 4 43E
Charlotte USA **79** L4 35 03N 80 50W
Charters Towers France **72** E3 20 04S 146 16E
Chartres France **48** D4 48 27N 1 30E
Châteauroux France **48** D3 46 49N 1 41E
Chatham England **39** E2 51 23N 0 32E
Cheadle England **34** C2 53 24N 2 13W
Cheb Czech Republic **49** G4 50 08N 12 28E
Cheboksary Russia **56** E3 56 08N 47 12E
Chelmsford England **39** E2 51 44N 0 28E
Cheltenham England **38** B2 51 54N 2 04W
Chelyabinsk Russia **56** G3 55 12N 61 25E
Chemnitz Germany **49** G4 50 50N 12 55E
Chengdu China **59** B3 30 37N 104 06E
Chennai India **58** D2 13 05N 80 18E
Chepstow Wales **37** E1 51 39N 2 41W
Cher *r.* France **48** D3 47 15N 1 30E
Cherbourg France **48** C4 49 38N 1 37W
Chernivtsi Ukraine **42** M5 48 19N 25 52E
Cherwell *r.* England **38** C2 51 55N 1 20W
Cheshire *co.* England **34** C2 53 30N 2 30W
Cheshunt England **38** D2 51 43N 0 02W
Chester England **34** C2 53 12N 2 54W
Chesterfield England **37** F3 53 15N 1 25W
Chester-le-Street England **34** D3 54 52N 1 34W
Cheviot Hills England/Scotland **33** G3 55 25N 2 20W
Cheviot, The *sum.* England **33** G3 55 29N 2 20W
Cheyenne USA **79** J4 41 08N 104 50W
Chiang Mai Thailand **60** A4 18 48N 98 59E
Chiba Japan **59** N2 35 38N 140 07E
Chicago USA **79** L5 41 50N 87 45W
Chichester England **38** D1 50 50N 0 48W
Chiclayo Peru **84** B7 6 47S 79 47W
Chicoutimi Canada **79** M5 48 26N 71 06W
Chieti Italy **51** E5 42 21N 14 10E
Chihuahua Mexico **79** J3 28 40N 106 06W
CHILE **84/85** B6/C2
Chillán Chile **85** B4 36 37S 72 10W
Chiloé Island Chile **85** B3 42 30S 74 00W
Chiltern Hills England **38** C2/D2 51 40N 1 00W
Chimbote Peru **84** B7 9 04S 78 34W
CHINA **59**
Chippenham England **38** B2 51 28N 2 07W
Chipping Norton England **38** C2 51 56N 1 32W
Chipping Sodbury England **38** B2 51 33N 2 24W
Chişinău Moldova **42** M5 47 00N 28 50E
Chita Russia **57** M3 52 03N 113 35E
Chittagong Bangladesh **58** E3 22 20N 91 48E
Cholet France **48** C3 47 04N 0 53W
Chongjin North Korea **59** D4 41 50N 129 55E
Chongqing China **59** B2 29 30N 106 35E
Chorley England **34** C2 53 40N 2 38W
Christchurch England **41** F3 50 44N 1 45W
Christchurch New Zealand **73** C2 43 32S 172 38E
Christmas Island Indian Ocean **60** B1 10 28S 105 35E
Churchill Canada **78** K6 58 45N 94 00W
Churchill Falls *tn.* Canada **78** N6 53 35N 64 00W
Cincinnati USA **79** L4 39 10N 84 30W
Cirencester England **38** C2 51 44N 1 59W
Citlatépetl *mt.* Mexico **79** K2 19 00N 97 18W
City of Edinburgh *u.a.* Scotland **33** F3 55 55N 3 15W
City of Kingston upon Hull *u.a.* England **35** E2 53 45N 0 25W
Ciudad Bolívar Venezuela **84** C8 8 06N 63 36W
Ciudadela Balearic Islands **50** G3 40 00N 3 50E
Ciudad Guayana Venezuela **84** C8 8 22N 62 37W
Ciudad Juárez Mexico **79** J4 31 42N 106 29W
Ciudad Obregón Mexico **79** J3 27 28N 109 59W
Ciudad Rodrigo Spain **50** B4 40 36N 6 33W
Clachan Scotland **32** D3 55 45N 5 34W
Clackmannanshire *u.a.* Scotland **33** F4 56 10N 3 40W
Clacton on Sea England **39** F2 51 48N 1 09E
Claonaig Scotland **32** D3 55 46N 5 22W
Cleethorpes England **35** E2 53 34N 0 02W

Clermont-Ferrand France **48** E2 45 47N 3 05E
Clevedon England **38** B2 51 27N 2 51W
Cleveland Hills England **35** D3 54 25N 1 15W
Clisham *mt.* Scotland **30** C2 57 58N 6 50W
Clitheroe England **34** C2 53 53N 2 23W
Clones Republic of Ireland **32** B2 54 11N 7 15W
Cluj-Napoca Romania **42** L5 46 47N 23 37E
Clutha *r.* New Zealand **73** B1 46 00S 169 30E
Clwyd *r.* Wales **36** D3 53 15N 3 25W
Clyde *r.* Scotland **33** F3 55 45N 3 55W
Clydebank Scotland **33** E3 55 54N 4 24W
Coalisland Northern Ireland **32** C2 54 32N 6 42W
Coalville England **37** F2 52 44N 1 20W
Coast Mountains Canada **78** G6 58 10N 132 40W
Coatbridge Scotland **33** E3 55 52N 4 01W
Coburg Germany **49** F4 50 15N 10 58E
Cochabamba Bolivia **84** C6 17 26S 66 10W
Cochin India **58** C1 9 56N 76 15E
Cockermouth England **34** B3 54 40N 3 21W
Cognac France **48** C2 45 42N 0 19W
Coimbatore India **58** C2 11 00N 76 57E
Coimbra Portugal **50** A4 40 12N 8 25W
Colchester England **39** E2 51 54N 0 54E
Coldstream *tn.* Scotland **33** G3 55 39N 2 15W
Coleraine Northern Ireland **32** C3 55 08N 6 40W
Coleraine *district* Northern Ireland **32** C3 55 00N 6 40W
Coll *i.* Scotland **30** C1 56 40N 6 35W
Colmar France **48** G4 48 05N 7 21E
Colne France **34** C2 53 52N 2 09W
Colne *r.* England **39** E2 51 53N 0 45E
Cologne Germany **49** D4 50 56N 6 57E
COLOMBIA **84** B8
Colombo Sri Lanka **58** C1 6 55N 79 52E
Colonsay *i.* Scotland **32** C4 56 05N 6 15W
Colorado *r.* Argentina **85** C4 37 30S 69 00W
Colorado *r.* USA/Mexico **79** J4 38 00N 109 00W
Colorado *state* USA **79** J4 39 00N 106 00W
Columbia USA **79** L4 34 00N 81 00W
Columbia *r.* Canada/USA **78** H6 50 30N 119 00W
Columbus USA **79** L4 39 59N 83 03W
Colwyn Bay *tn.* Wales **36** D3 53 18N 3 43W
Communism Peak Tajikistan **56** H1 38 59N 72 01E
Como Italy **51** B6 45 48N 9 05E
Comodoro Rivadavia Argentina **85** C3 45 50S 67 30W
Como, Lake Italy **51** B6/7 46 00N 9 00E
COMOROS **67** E3
Conakry Guinea **66** A3 9 30N 13 43W
Concepción Chile **85** B4 36 50S 73 03W
Conejara *i.* Balearic Islands **50** G3 39 11N 2 58E
CONGO **66** D2
Congo *r.* Africa **67** B4 3 00S 16 00E
CONGO DEMOCRATIC REPUBLIC **67** C4
Coniston Water *l.* England **34** B3 54 20N 3 05W
Connecticut *state* USA **79** M5 41 00N 73 00W
Consett England **34** D3 54 51N 1 49W
Constance, Lake Europe **49** E2 47 40N 9 30E
Constanţa Romania **42** M4 44 12N 28 40E
Constantine Algeria **66** C6 36 22N 6 40E
Conwy Wales **36** D3 53 17N 3 50W
Conwy *r.* Wales **36** D3 53 20N 3 55W
Conwy *u.a.* Wales **36** D3 53 14N 3 50W
Cook Islands Pacific Ocean **4** 19 30S 159 50W
Cook, Mount New Zealand **73** C2 43 36S 170 09E
Cookstown Northern Ireland **32** C2 54 39N 6 45W
Cookstown *district* Northern Ireland **32** C2 54 35N 6 55W
Cook Strait New Zealand **73** C2 41 24S 174 36E
Cooktown Australia **72** E4 15 28S 145 15E
Cooper Creek Australia **72** E3 26 10S 142 20E
Copenhagen Denmark **42** J7 55 43N 12 34E
Coquet *r.* England **33** H3 55 20N 1 50W
Coral Sea Pacific Ocean **72** F4 15 00S 154 00E
Corby England **38** D3 52 49N 0 32W
Córdoba Argentina **85** C4 31 25S 64 11W
Córdoba Spain **50** C2 37 53N 4 46W
Corfu *i.* Greece **42** K3 39 00N 19 00E
Cornwall *co.* England **40** B3/C3 50 25N 4 50W
Corpus Christi USA **79** K3 27 47N 97 26W
Corrientes Argentina **85** D5 27 30S 58 48W
Corsewall Point Scotland **32** D3 55 00N 5 10W
Corsica *i.* France **51** B5 42 00N 9 00E
Cosenza Italy **51** F3 39 17N 16 16E
Costa Brava *geog. reg.* Spain **50** G4/5 41 40N 3 50E
Costa del Sol *geog. reg.* Spain **50** C2/D2 36 40N 4 40W
COSTA RICA **79** L1
CÔTE D'IVOIRE **66** B3
Cotonou Benin **66** C2 6 24N 2 31E
Cotopaxi *vol.* Ecuador **84** B7 0 40S 78 28W
Cotswold Hills England **38** B2/C2 51 40N 2 10W
Cottbus Germany **49** H4 51 43N 14 21E
Coupar Angus Scotland **31** F1 56 33N 3 17W
Coutainville France **41** F2 49 03N 1 34W
Coutances France **41** F2 49 03N 1 29W
Coventry England **37** F2 52 25N 1 30W
Cowes England **38** C1 50 45N 1 18W
Craigavon Northern Ireland **32** C2 54 28N 6 25W
Craigavon *district* Northern Ireland **32** C2 54 30N 6 30W
Craighouse Scotland **32** D3 55 51N 5 57W
Craignure Scotland **30** D1 56 28N 5 42W
Crail Scotland **33** G4 56 16N 2 38W
Craiova Romania **42** L4 44 18N 23 47E
Cramlington England **33** H3 55 05N 1 35W
Crawley England **38** D2 51 07N 0 12W
Crediton England **40** D3 50 47N 3 39W
Cree *r.* Scotland **33** E3 55 00N 4 30W
Creeslough Republic of Ireland **32** B3 55 07N 7 55W
Cremona Italy **51** C6 45 08N 10 01E
Crete *i.* Greece **42** M3 35 00N 25 00E
Crewe England **34** C2 53 05N 2 27W
Crewkerne England **41** E3 50 53N 2 48W
Crianlarich Scotland **31** E1 56 23N 4 37W
Crieff Scotland **31** F1 56 23N 3 52W
Crimea *p.* Ukraine **43** N5 46 00N 34 00E
CROATIA **42** K5
Cromarty Scotland **31** E2 57 40N 4 02W
Cromarty Firth *est.* Scotland **31** E2 57 40N 4 20W
Cromer England **39** F3 52 56N 1 18E

Cross Fell *sum.* England **34** C3 54 43N 2 29W
Crossmaglen Northern Ireland **32** C2 54 05N 6 37W
Crotone Italy **51** F3 39 05N 17 08E
Croydon England **38** D2 51 23N 0 06W
Crumlin Northern Ireland **32** C2 54 37N 6 14W
Cruzeiro do Sul Brazil **84** B7 7 40S 72 39W
Cuanza *r.* Angola **67** B4 9 40S 15 00E
CUBA **79** M3
Cubango *r.* Southern Africa **67** B3 17 00S 18 00E
Cúcuta Colombia **84** B8 7 55N 72 31W
Cuenca Ecuador **84** B7 2 54S 79 00W
Cuiabá Brazil **84** D6 15 32S 56 05W
Cuillin Hills Scotland **30** C2 57 15N 6 15W
Culiacán Mexico **79** J3 24 50N 107 23W
Cullen Scotland **31** G2 57 41N 2 49W
Cullompton England **40** D3 50 52N 3 24W
Cumaná Venezuela **84** C9 10 29N 64 12W
Cumbernauld Scotland **33** F3 55 57N 4 00W
Cumbria *co.* England **34** B3/C3 54 30N 2 50W
Cumnock Scotland **33** E3 55 27N 4 16W
Cunene *r.* Angola/Namibia **67** B3 17 00S 15 00E
Cuneo Italy **51** A6 44 24N 7 33E
Cunnamulla Australia **72** E3 28 03S 145 43E
Cupar Scotland **33** F4 56 19N 3 01W
Curaçao *i.* Lesser Antilles **79** N2 12 20N 68 20W
Curitiba Brazil **85** E5 25 25S 49 25W
Cuttack India **58** D3 20 26N 85 56E
Cuzco Peru **84** B6 13 32S 1 57W
Cwmbran Wales **36** D1 51 39N 3 00W
Cyclades *is.* Greece **42** L3/L4 36 00N 25 00E
CYPRUS **43** N2
CZECH REPUBLIC **42** K5
Częstochowa Poland **42** K6 50 49N 19 07E

D

Daer Reservoir Scotland **33** F3 55 25N 3 36W
Dakar Senegal **66** A4 14 38N 17 27W
Dalbeattie Scotland **33** F2 54 56N 3 49W
Dalian China **59** D3 38 53N 121 37E
Dallas USA **79** K4 32 47N 96 48W
Dalmally Scotland **31** E1 56 24N 4 58W
Dalton-in-Furness England **34** B3 54 09N 3 11W
Daly *r.* Australia **72** D4 13 30S 130 30E
Damascus Syria **61** D4 33 30N 36 19E
Dammam Saudi Arabia **61** F3 26 25N 50 06E
Da Nang Vietnam **60** B4 16 04N 108 14E
Danube *r.* Europe **42** L4 44 00N 25 00E
Dar es Salaam Tanzania **67** D4 6 51S 39 18E
Darfur *mts.* Sudan **61** C2 12 40N 24 20E
Dargaville New Zealand **73** C3 35 56S 173 52E
Darling *r.* Australia **72** E2 31 00S 144 00E
Darlington England **34** D3 54 31N 1 34W
Darlington *u.a.* England **34** D3 54 35N 1 34W
Darmstadt Germany **49** E3 49 52N 8 39E
Darnah Libya **42** L2 32 46N 22 39E
Dart *r.* England **40** D3 50 30N 3 50W
Dartford England **39** E2 51 27N 0 13E
Dartmoor *geog. reg.* England **40** D3 50 35N 3 50W
Dartmouth England **40** D3 50 21N 3 35W
Darvel Scotland **33** E3 55 37N 4 18W
Darwin Australia **72** D4 12 24S 130 52E
Daugava *r.* Belarus **42** M7 55 00N 29 00E
Daugavpils Latvia **42** M7 55 52N 26 31E
Dauphine *geog. reg.* France **48** F2 45 00N 5 00E
Davao The Philippines **60** D3 7 05N 125 38E
Daventry England **38** C3 52 16N 1 09W
Dawlish England **40** D3 50 35N 3 28W
Dawson Canada **78** F7 64 04N 139 24W
Deal England **39** F2 51 14N 1 24E
Debrecen Hungary **42** L5 47 30N 21 37E
Deccan *plat.* India **58** C2 18 00N 78 00E
Dee *r.* Aberdeenshire/Aberdeen City Scotland **31** G2 57 05N 2 10W
Dee *r.* Dumfries and Galloway Scotland **33** E2/F2 54 55N 4 00W
Dee *r.* England/Wales **36** D3 53 20N 3 10W
Deeps, The *b.* Scotland **29** D3 60 10N 1 25W
Delaware *state* USA **79** M4 39 00N 75 00W
Delhi India **58** C3 28 40N 77 14E
Demavand, Mount Iran **61** F4 35 56N 52 08E
Denbigh Wales **36** D3 53 11N 3 25W
Denbighshire *u.a.* Wales **36** D3 53 10N 3 10W
DENMARK **42** H7
Denver USA **79** J4 39 45N 105 00W
De Panne Belgium **39** G2 51 06N 2 35E
Derby Australia **72** C4 17 18S 123 37E
Derby England **37** F2 52 55N 1 30W
Derby City *u.a.* England **37** F2 52 53N 1 30W
Derbyshire *co.* England **37** F3 53 10N 1 30W
Derg *r.* Northern Ireland **32** B2 54 40N 7 45E
Derwent *r.* Cumbria England **34** B3 54 40N 3 30W
Derwent *r.* Derbyshire England **37** F3 53 20N 1 40W
Derwent *r.* Northumberland/Durham/Tyne and Wear England **34** D3 54 50N 2 05W
Derwent *r.* North Yorkshire/York/East Riding of Yorkshire England **35** E2 53 50N 0 55W
Derwent Water *l.* England **34** B3 54 35N 3 09W
Des Moines USA **79** K5 41 35N 93 35W
Dessau Germany **49** G4 51 51N 12 15E
Detroit USA **79** L5 42 23N 83 05W
Deveron *r.* Scotland **31** G2 57 35N 2 40W
Devizes England **38** C2 51 22N 1 59W
Devon *co.* England **40** C3/D3 50 45N 3 50W
Devon Island Canada **78** L8 75 47N 88 00W
Dewsbury England **34** D2 53 42N 1 37W
Dezful Iran **61** F4 32 23N 48 28E
Dhaka Bangladesh **58** E3 23 42N 90 22E
Dibrugarh India **58** E3 27 29N 94 56E
Didcot England **38** C2 51 37N 1 15W
Dieppe France **48** D4 49 55N 1 05E
Dijon France **48** F3 47 20N 5 02E
Dili East Timor **60** D2 8 33S 125 34E
Dingwall Scotland **31** E2 57 36N 4 29W
Dire Dawa Ethiopia **61** E1 9 35N 41 50E
Diss England **39** F3 52 23N 1 06W
Diyarbakir Turkey **43** Q3 37 55N 40 14E
Djelfa Algeria **42** J4 34 43N 3 14E
DJIBOUTI **61** E2
Djibouti Djibouti **61** E2 11 35N 43 11E

Dnieper r. Ukraine/Belarus **56** D3 51 00N 30 00E
Dniester r. Europe **42** M5 48 00N 27 00E
Dnipropetrovsk Ukraine **56** D2 48 29N 35 00E
Dodoma Tanzania **67** D4 6 10S 35 40E
Doha Qatar **61** F3 25 15N 5 13E
Dolgellau Wales **36** D2 52 44N 3 53W
Dolomites mts. Italy **49** F2 46 00N 12 00E
DOMINICA **79** N2
DOMINICAN REPUBLIC **79** M2
Don r. Russia **56** E3 50 00N 40 30E
Don r. Scotland **31** G2 57 15N 2 10W
Donaghadee Northern Ireland **32** D2 54 39N 5 33W
Doncaster England **35** D2 53 32N 1 07W
Donegal Republic of Ireland **32** A2 54 39N 8 07W
Donegal co. Republic of Ireland **32** A2/B2 54 55N 8 00W
Donetsk Ukraine **56** D2 48 00N 37 50E
Dongting Hu l. China **59** C2 29 00N 113 30E
Doon r. Scotland **33** E3 55 20N 4 30W
Dorchester England **41** E3 50 43N 2 26W
Dordogne r. France **48** D2 45 00N 1 30E
Dordrecht Netherlands **49** C4 51 48N 4 40E
Dorking England **38** D2 51 14N 0 20W
Dornoch Scotland **31** E2 57 52N 4 02W
Dornoch Firth est. Scotland **31** E2 57 55N 3 55W
Dorset co. England **41** E3 50 50N 2 20W
Dortmund Germany **49** D4 51 32N 7 27E
Douai France **48** E5 50 22N 3 05E
Douala Cameroon **66** C3 4 04N 9 43E
Douglas Isle of Man British Isles **33** E2 54 09N 4 29W
Douro r. Spain/Portugal **50** B4 41 30N 6 30W
Dove r. England **37** E3 53 10N 1 50W
Dover England **39** F2 51 08N 1 19E
Dover, Strait of English Channel **39** F1/2 51 00N 1 20W
Down district Northern Ireland **32** D2 54 25N 5 55W
Downham Market England **39** E3 52 36N 0 23E
Downpatrick Northern Ireland **32** D2 54 20N 5 43W
Drakensberg mts. Republic of South Africa **67** C2 30 00S 28 00E
Drava r. Europe **49** G2 46 45N 13 30E
Dresden Germany **49** G4 51 03N 13 45E
Droitwich England **37** E2 52 16N 2 09W
Dromore Northern Ireland **32** C2 54 25N 6 09W
Dronning Maud Land geog. reg. Antarctica **86** 73 00S 10 00E
Drummore Scotland **32** E2 54 42N 4 54W
Drumnadrochit Scotland **31** E2 57 20N 4 30W
Dubai United Arab Emirates **61** F3 25 14N 55 17E
Dubbo Australia **72** E2 32 15S 148 38E
Dublin Republic of Ireland **36** A3 53 20N 6 15W
Dudley England **37** E2 52 30N 2 05W
Dufftown Scotland **31** F2 57 26N 3 08W
Duisburg Germany **49** D4 51 26N 6 45E
Duluth USA **79** K5 46 45N 92 10W
Dumbarton Scotland **33** E3 55 57N 4 35W
Dumfries Scotland **33** F3 55 04N 3 37W
Dumfries and Galloway u.a. Scotland **33** E3/F3 55 10N 4 00W
Dunbar Scotland **33** G4 56 00N 2 31W
Dunblane Scotland **33** F4 56 12N 3 59W
Duncansby Head c. Scotland **29** E3 58 39N 3 02W
Dundalk Republic of Ireland **32** C2 54 01N 6 25W
Dundee Scotland **31** G1 56 28N 3 00W
Dundee City u.a. Scotland **31** F1/G1 56 30N 2 55W
Dunedin New Zealand **73** C1 45 53S 170 30E
Dunfermline Scotland **33** F4 56 04N 3 29W
Dungannon Northern Ireland **32** C2 54 31N 6 46W
Dungannon district Northern Ireland **32** B2/C2 54 30N 7 00W
Dungeness c. England **39** E1 50 55N 0 58E
Dungiven Northern Ireland **32** C2 54 55N 6 55W
Dunkerque France **48** E5 51 02N 2 23E
Dunkery Beacon sum. England **40** D4 51 11N 3 35W
Dún Laoghaire Republic of Ireland **36** A3 53 17N 6 08W
Dunnett Head Scotland **29** B1 58 40N 3 25W
Dunoon Scotland **32** E3 55 57N 4 56W
Duns Scotland **33** G3 55 47N 2 20W
Dunstable England **38** D2 51 53N 0 32W
Dunvegan Scotland **30** C2 57 26N 6 35W
Durance r. France **48** F1 43 50N 5 15E
Durango Mexico **79** J3 24 01N 104 40W
Durban Republic of South Africa **67** D2 29 53S 31 00E
Durham England **34** D3 54 47N 1 34W
Durham co. England **34** C3/D3 54 40N 2 00W
Dursley England **37** E1 51 42N 2 21W
Dushanbe Tajikistan **56** G1 38 38N 68 51E
Düsseldorf Germany **49** D4 51 13N 6 47E
Dyce Scotland **31** G2 57 12N 2 11W
Dyfi r. Wales **36** D2 52 38N 3 50W

E

Ealing England **38** D2 51 31N 0 18W
Earn r. Scotland **31** F1 56 15N 3 40W
East Ayrshire u.a. Scotland **33** E3 55 35N 4 15W
Eastbourne England **39** E1 50 46N 0 17E
East Cape New Zealand **73** D3 37 41S 178 33E
East Dereham England **39** E3 52 41N 0 56E
East Dunbartonshire u.a. Scotland **33** E3 55 55N 4 15W
Eastern Rift Valley East Africa **67** D4/5 4 00N 36 00E
East Friesian Islands Netherlands **49** D5 53 00N 7 00E
East Kilbride Scotland **33** E3 55 46N 4 10W
Eastleigh England **38** C1 50 58N 1 22W
East London Republic of South Africa **67** C1 33 00S 27 54E
East Lothian u.a. Scotland **33** G3 55 56N 2 42W
East Renfrewshire u.a. Scotland **33** E3 55 45N 4 26W
East Riding of Yorkshire u.a. England **35** E3 53 54N 0 40W
East Sussex co. England **39** E1 50 55N 0 10E
EAST TIMOR **60** D2
Ebbw Vale Wales **36** D1 51 47N 3 12W
Ebro r. Spain **50** E4 42 00N 1 15W
Ech Cheliff Algeria **42** G3 36 05N 1 15E
ECUADOR **84** B7

Eday i. Scotland **29** C2 59 11N 2 47W
Eddrachillis Bay Scotland **30** D3 58 25N 5 15W
Eden r. Cumbria England **34** C3 54 45N 2 40W
Eden r. Surrey/Kent England **39** E2 51 10N 0 05E
Edinburgh Scotland **33** F3 55 57N 3 13W
Edmonton Canada **78** H6 53 34N 113 25W
Edward, Lake Congo Dem. Rep./Uganda **67** C4 0 30S 29 40E
EGYPT **61** C3
Eigg i. Scotland **30** C1 56 55N 6 10W
Eindhoven Netherlands **49** C4 51 26N 5 30E
Elba i. Italy **51** C5 42 00N 10 00E
El Bayadh Algeria **42** G2 33 40N 1 00E
Elbe r. Europe **49** F5 53 15N 11 00E
Elbrus, Mount Russia **43** G2 43 21N 42 29E
Elche Spain **50** E3 38 16N 0 41W
Eleuthera i. The Bahamas **79** M3 25 05N 76 30W
El Faiyum Egypt **61** D3 29 19N 30 50E
El Fasher Sudan **61** C2 13 37N 25 22E
El Gîza Egypt **61** D4 30 01N 31 12E
Elgol Scotland **30** C2 57 09N 6 06W
El Kharga Egypt **61** D3 25 27N 30 32E
Ellen r. England **34** B3 54 45N 3 25W
Ellesmere England **37** E2 52 55N 2 54W
Ellesmere Island Canada **78** L8 77 30N 82 30W
Ellesmere Port England **34** C2 53 17N 2 54W
Ellon Scotland **31** G2 57 22N 2 05W
Ellsworth Land geog. reg. Antarctica **86** 75 00S 80 00W
El Minya Egypt **61** D3 28 06N 30 45E
El Paso USA **79** J4 31 45N 106 30W
EL SALVADOR **79** L2
Elvas Portugal **50** B3 38 53N 7 10W
Ely England **39** E3 52 24N 0 16E
Emden Germany **49** D5 53 23N 7 13E
Emi Koussi mt. Chad **66** D4 19 52N 18 31E
Ems r. Germany **49** D5 53 00N 7 00E
Enard Bay Scotland **30** D3 58 10N 5 25W
Enfield England **38** D2 51 39N 0 05W
Enggano i. Indonesia **60** B2 5 20S 102 20E
England country United Kingdom **28** 53 00N 2 00W
English Channel United Kingdom/France **28** 50 00N 2 00W
Enniskillen Northern Ireland **32** B2 54 21N 7 38W
Enschede Netherlands **49** D5 52 13N 6 55E
Entebbe Uganda **67** D5 0 04N 32 27E
EQUATORIAL GUINEA **66** D3
Erebus, Mount Antarctica **86** 77 40S 167 20E
Erfurt Germany **49** F4 50 58N 11 02E
Erie, Lake Canada/USA **79** L5 42 15N 81 00W
Eriskay i. Scotland **30** B2 57 05N 7 10W
ERITREA **61** D2
Erlangen Germany **49** F3 49 36N 11 02E
Erne r. Northern Ireland/Republic of Ireland **32** B2 54 10N 7 30W
Errigal Mountain Republic of Ireland **32** A3 55 02N 8 07W
Erzurum Turkey **43** Q3 39 57N 41 17E
Esfahan Iran **61** F4 32 41N 51 41E
Esha Ness pt. Scotland **29** D3 60 29N 1 37W
Esk r. England **35** E3 54 30N 0 45W
Esk r. Scotland/England **33** F3 55 15N 3 05W
Eskisehir Turkey **43** N3 39 46N 30 30E
Esperance Australia **72** C2 33 51S 121 54E
Esquel Argentina **85** B3 42 55S 71 20W
Essen Germany **49** D4 51 27N 6 57E
Essex co. England **39** E2 51 46N 0 30E
ESTONIA **42** M7
ETHIOPIA **61** E1
Ethiopian Highlands Africa **61** D1 8 00N 37 00E
Etna, Mount Italy **51** E2 37 45N 15 00E
Etosha Pan salt l. Namibia **67** B3 18 30S 16 30E
Eyre, Lake Australia **72** D3 28 43S 137 11E
Euphrates r. Iraq/Syria/Turkey **61** D4/E4 35 00N 40 00E
Europa i. Mozambique Channel **67** E2 22 20S 40 20E
Everest, Mount China/Nepal **58** D3 27 59N 86 56E
Evesham England **37** E2 52 06N 1 56W
Evreux France **48** D4 49 03N 1 11E
Evvoia i. Greece **42** L3 38 00N 24 00E
Exe r. England **40** D4 51 10N 3 35W
Exeter England **40** D3 50 43N 3 31W
Exmoor geog. reg. England **40** D4 51 08N 3 40W
Exmouth England **40** D3 50 37N 3 25W
Eyemouth Scotland **33** G3 55 25N 2 06W
Eye Peninsula Scotland **30** C3 58 10N 6 10W

F

Faroe Islands Atlantic Ocean **42** E8 62 00N 7 00W
Fairbanks USA **78** E7 64 50N 147 50W
Fair Head c. Northern Ireland **32** C3 55 15N 6 10W
Fair Isle i. Shetland Islands Scotland **29** D2 59 32N 1 38W
Faisalabad Pakistan **58** C4 31 25N 73 09E
Fakenham England **39** E3 52 50N 0 51E
Fal r. England **40** C3 50 15N 4 55W
Falkirk Scotland **33** F3 55 59N 3 48W
Falkirk u.a. Scotland **33** F3/4 56 05N 3 53W
Falkland Islands South Atlantic Ocean **85** D2 52 30S 60 00W
Falmouth England **40** B3 50 08N 5 04W
Falster r. Denmark **49** F6/G6 54 00N 12 00E
Fareham England **38** C1 50 51N 1 10W
Farewell, Cape Greenland **78** Q6 60 00N 44 00W
Farewell, Cape New Zealand **73** C2 40 30S 172 41E
Faringdon England **38** C2 51 40N 1 35W
Farnborough England **38** C2 51 17N 0 46W
Farnham England **38** D2 51 13N 0 49W
Faro Portugal **50** B2 37 01N 7 56W
Fawley England **38** C1 50 49N 1 20W
FEDERATED STATES OF MICRONESIA **5**
Fehmarn i. Germany **49** F6 54 00N 11 00E
Feilding New Zealand **73** D2 40 14S 175 34E
Feira de Santana Brazil **84** F6 12 17S 38 53W
Felixstowe England **39** F2 51 58N 1 20E
Fens, The geog. reg. England **35** E1/F1 52 45N 0 05E
Fergana Uzbekistan **56** H2 40 23N 71 19E

Fermanagh district Northern Ireland **32** B2 54 25N 7 45W
Ferrara Italy **51** C6 44 50N 11 38E
Fès Morocco **66** B6 34 05N 5 00W
Fethaland, Point of Scotland **29** D3 60 35N 1 18W
Fetlar i. Scotland **29** E3 60 37N 0 52W
Fife u.a. Scotland **33** F4/G4 56 10N 3 00W
FIJI **5**
Filchner Ice Shelf Antarctica **86** 80 00S 37 00W
Finke r. Australia **72** D3 25 00S 135 00E
FINLAND **42** M8
Finland, Gulf of Finland/Russia **42** L7 59 40N 23 30E
Finn r. Northern Ireland/Republic of Ireland **32** B2 54 50N 7 50W
Fionnphort Scotland **30** C1 56 19N 6 23W
Firth of Clyde est. Scotland **32** E3 55 45N 5 00W
Firth of Forth est. Scotland **33** F4/G4 56 05N 3 00W
Firth of Lorn est. Scotland **30** D1 56 15N 6 00W
Firth of Tay est. Scotland **31** F1 56 25N 3 00W
Fishguard Wales **36** C1 51 59N 4 59W
Fitzroy r. Australia **72** C4 17 00S 126 00E
Flamborough Head c. England **35** D3 54 06N 0 04W
Fleetwood England **34** B2 53 56N 3 01W
Flinders r. Australia **72** E3/4 20 00S 142 00E
Flinders Range Australia **72** D2 31 20S 138 45E
Flint Wales **36** D3 53 15N 3 07W
Flintshire u.a. Wales **36** D3 53 25N 3 10W
Florence Italy **51** C5 43 47N 11 15E
Flores i. Indonesia **60** D2 8 30S 121 00E
Flores Sea Indonesia **60** D2 7 00S 119 00E
Florianópolis Brazil **85** E5 27 35S 48 31W
Florida state USA **79** L3 28 00N 82 00W
Florida, Straits of sd. Cuba/USA **79** L3 24 00N 81 00W
Fly r. Papua New Guinea **72** E5 8 00S 142 00E
Fochabers Scotland **31** F2 57 37N 3 05W
Fóggia Italy **51** E4 41 28N 15 33E
Folkestone England **39** F2 51 05N 1 11E
Fontainebleau France **48** E4 48 24N 2 42E
Forel, Mount Greenland **86** 67 00N 37 00W
Forest of Bowland hills England **34** C2 54 00N 2 35W
Forfar Scotland **31** G1 56 38N 2 54W
Forlì Italy **51** D6 44 13N 12 02E
Formby England **34** B2 53 34N 3 05W
Formentera i. Balearic Islands **50** F3 38 41N 1 30E
Fornells Balearic Islands **50** H3 40 04N 4 07E
Forres Scotland **31** F2 57 37N 3 38W
Fortaleza Brazil **84** F7 3 45S 38 35W
Fort Augustus Scotland **31** E2 57 09N 4 41W
Forth r. Scotland **33** E4 56 10N 4 10W
Fort McMurray Canada **78** H6 56 45N 111 27W
Fort Peck Reservoir USA **79** J5 48 01N 106 28W
Fort Simpson Canada **78** G7 61 46N 121 15W
Fort William Scotland **30** D1 56 49N 5 07W
Fort Worth USA **79** K4 32 45N 97 20W
Foula i. Scotland **29** C3 60 08N 2 05W
Foveaux Strait New Zealand **73** B1 46 42S 167 59E
Fowey England **40** C3 50 20N 4 38W
Fowey r. England **40** C3 50 25N 4 30W
Foyle r. Northern Ireland/Republic of Ireland **32** B2 54 58N 7 35W
FRANCE **48**
Frankfurt Germany **49** E4 50 06N 8 41E
Franz Josef Land is. Russia **57** G6 80 00N 50 00E
Fraser r. Canada **78** G6 52 00N 122 00W
Fraserburgh Scotland **31** G2 57 42N 2 00W
Freetown Sierra Leone **66** A3 8 30N 13 17W
Freiburg Germany **49** D3 48 00N 7 52E
Fréjus France **48** G1 43 26N 6 44E
Fremantle Australia **72** B2 32 03S 115 46E
French Guiana territory France **84** D8 5 00N 53 00W
French Polynesia is. Pacific Ocean **4** 21 00S 150 00W
Fresno USA **79** H4 36 41N 119 47W
Frome England **41** E4 51 14N 2 20W
Frome r. England **41** E3 50 40N 2 20W
Fuji, Mount Japan **59** M2 35 23N 138 42E
Fukui Japan **59** M2 36 04N 136 12E
Fukuoka Japan **59** L1 33 39N 130 21E
Fukushima Japan **59** N2 37 44N 140 28E
Fukuyama Japan **59** L1 34 29N 133 21E
Furnace Scotland **32** D4 56 09N 5 12W
Fushun China **59** D4 41 50N 123 54E
Fuzhou China **59** C2 26 09N 119 17E
Fyn i. Denmark **49** F6 55 30N 10 00E
FYROM see MACEDONIA, FORMER YUGOSLAV REPUBLIC OF

G

Gabès Tunisia **42** J2 33 52N 10 06E
GABON **66** D2
Gaborone Botswana **67** C2 24 45S 25 55E
Gafsa Tunisia **42** H2 34 28N 8 43E
Gainsborough England **35** E2 53 24N 0 48W
Gairloch tn. Scotland **30** D2 57 42N 5 40W
Galapagos Islands Pacific Ocean **4** 12 00S 87 00W
Galashiels Scotland **33** G3 55 37N 2 49W
Galaţi Romania **42** M5 45 27N 28 02E
Gallipoli Italy **51** G4 40 03N 17 59E
GAMBIA, THE **66** A4
Gänca Azerbaijan **56** E2 40 39N 46 20E
Gander Canada **79** P5 49 00N 54 15W
Ganges r. India/Bangladesh **58** D3 24 00N 88 00E
Gap France **48** G2 44 33N 6 05E
Garda, Lake Italy **51** C6 45 00N 10 00E
Garelochhead Scotland **32** E4 56 05N 4 50W
Garonne r. France **48** D2 44 45N 0 15E
Gatehouse of Fleet Scotland **33** E2 54 53N 4 11W
Gateshead England **34** D3 54 58N 1 35W
Gaziantep Turkey **43** P3 37 04N 37 21E
Gdańsk Poland **42** K6 54 22N 18 41E
Gdynia Poland **42** K6 54 31N 18 30E
Geelong Australia **72** E2 38 08S 144 20E
Gelsenkirchen Germany **49** D4 51 30N 7 05E
Geneva Switzerland **49** D2 46 13N 6 09E
Geneva, Lake France/Switzerland **49** D2 46 20N 6 20E
Genoa Italy **51** B6 44 24N 8 56E
Gent Belgium **49** B4 51 02N 3 42E
Georgetown Guyana **84** D8 6 46N 58 10W
George Town Malaysia **60** B3 5 25N 100 20E
GEORGIA **56** E2

Georgia state USA **79** L4 33 00N 83 00W
Gera Germany **49** G4 50 51N 12 11E
Geraldton Australia **72** B3 28 46S 114 36E
GERMANY **49**
GHANA **66** B3
Gibraltar territory UK **50** C2 36 09N 5 21W
Gibraltar, Strait of Spain/Morocco **50** C1 35 58N 5 30W
Gibson Desert Australia **72** C3 24 00S 122 00E
Giessen Germany **49** E4 50 35N 8 42E
Gifu Japan **59** M2 35 27N 136 46E
Gigha i. Scotland **32** D3 55 40N 5 45W
Gijón Spain **50** C5 43 32N 5 40W
Gillingham England **39** E2 51 24N 0 33E
Gironde r. France **48** C2 45 30N 1 00W
Girvan Scotland **32** E3 55 15N 4 51W
Gisborne New Zealand **73** D3 38 40S 178 01E
Glasgow Scotland **33** E3 55 53N 4 15W
Glasgow City u.a. Scotland **33** E3 55 50N 4 20W
Glastonbury England **41** E4 51 09N 2 43W
Glenluce Scotland **32** E2 54 53N 4 49W
Glenrothes Scotland **33** F4 56 12N 3 10W
Gloucester England **38** B2 51 53N 2 14W
Gloucestershire co. England **38** B2 51 50N 2 20W
Goat Fell mt. Scotland **32** D3 55 39N 5 11W
Gobi Desert Mongolia **57** L2 48 30N 100 00E
Godavari r. India **58** C2/D2 19 00N 80 00E
Godthåb see Nuuk
Goiânia Brazil **84** E6 16 43S 49 18W
Gold Coast tn. Australia **72** F3 28 04S 153 25E
Gongga Shan mt. China **59** B2 29 34N 101 53E
Good Hope, Cape of Republic of South Africa **67** B1 34 30S 19 00E
Goole England **35** E2 53 42N 0 52W
Gore New Zealand **73** B1 46 06S 168 56E
Goroka Papua New Guinea **72** E5 6 02S 145 22E
Göteborg Sweden **42** J7 57 45N 12 00E
Gotland i. Sweden **42** K7 57 30N 18 40E
Göttingen Germany **49** E4 51 32N 9 57E
Governador Valadares Brazil **84** E6 18 51S 41 57W
Gower p. Wales **36** C1 51 35N 4 10W
Grafton Australia **72** F3 29 40S 152 58E
Grampian Mountains Scotland **31** E1/F1 56 45N 4 00W
Granada Spain **50** D2 37 10N 3 35W
Grand Bahama i. The Bahamas **79** M3 27 00N 78 00W
Grand Canyon USA **79** H4 36 04N 112 07W
Grand Rapids tn. USA **79** L5 42 57N 86 40W
Grangemouth Scotland **33** F4 56 01N 3 44W
Grantham England **35** E1 52 55N 0 39W
Grantown-on-Spey Scotland **31** F2 57 20N 3 58W
Gravesend England **39** E2 51 27N 0 24E
Grays England **39** E2 51 29N 0 20E
Graz Austria **42** K5 47 05N 15 22E
Great Abaco i. The Bahamas **79** M3 26 40N 77 00W
Great Australian Bight b. Australia **72** C2 32 31S 130 00E
Great Barrier Island New Zealand **73** D3 36 13S 175 24E
Great Barrier Reef Australia **72** E4/F3 15 00S 146 00E
Great Bear Lake Canada **78** H7 66 00N 120 00W
Great Dividing Range mts. Australia **72** E2/F3 35 00N 148 00E
Great Driffield England **35** E3 54 01N 0 26W
Greater Antilles is. West Indies **79** M2/N2 19 00N 78 00W
Greater London co. England **38** D2 51 30N 0 10W
Greater Manchester co. England **34** C2 53 30N 2 15W
Great Inagua i. The Bahamas **79** M3 21 40N 73 00W
Great Karoo mts. Republic of South Africa **67** C1 32 30S 22 30E
Great Malvern England **37** E2 52 07N 2 19W
Great Ouse r. England **39** E3 52 40N 0 20E
Great St. Bernard Pass France/Italy **48** G2 45 50N 7 10E
Great Salt Lake USA **79** H5 41 10N 112 40W
Great Sandy Desert Australia **72** C3 21 00S 121 00E
Great Slave Lake Canada **78** H7 62 00N 114 00W
Great Stour r. England **39** F2 51 20N 1 10E
Great Torrington England **40** C3 51 57N 4 09W
Great Victoria Desert Australia **72** C3/D3 28 00S 129 00E
Great Wall China **59** B3 39 00N 109 00E
Great Whernside sum. England **34** D3 54 10N 1 59W
Great Yarmouth England **39** F3 52 37N 1 44E
GREECE **42** L3
GREENLAND **78** Q8
Greenock Scotland **33** E3 55 57N 4 45W
Greenwich England **38** D2 51 29N 0 00
GRENADA **79** N2
Grenoble France **48** F2 45 11N 5 43E
Greta r. England **34** C3 54 10N 2 30W
Greymouth New Zealand **73** C2 42 27S 171 12E
Greystones Republic of Ireland **36** A3 53 09N 6 04W
Grimsby England **35** E2 53 35N 0 05W
Groningen Netherlands **49** D5 53 13N 6 35E
Grosseto Italy **51** C5 42 46N 11 07E
Gross Glockner mt. Austria **49** G2 47 05N 12 44E
Groznyy Russia **56** E2 43 21N 45 42E
Guadalajara Mexico **79** J3 20 40N 103 20W
Guadalquivir r. Spain **50** C2 37 45N 5 30W
Guadalupe i. Mexico **79** H3 29 00N 118 24W
Guadeloupe i. Lesser Antilles **79** N2 16 30N 61 30W
Guadiana r. Spain/Portugal **50** B3 39 00N 5 30W
GUAM **5**
Guangzhou China **59** C2 23 08N 113 20E
Guantánamo Cuba **79** M3 20 09N 75 14W
Guarda Portugal **50** B4 40 32N 7 17W
GUATEMALA **79** K2
Guatemala Guatemala **79** K2 14 38N 90 22W
Guayaquil Ecuador **84** B7 2 13S 79 54W
Guernsey i. Channel Islands British Isles **41** E2 46 27N 2 35W
Guildford England **38** D2 51 14N 0 35W
GUINEA **66** A4
GUINEA-BISSAU **66** A4
Guinea, Gulf of Atlantic Ocean **66** C3 2 00N 2 30E
Guisborough England **35** D3 54 32N 1 04W
Guiyang China **59** B2 26 35N 106 40E

Gulf, The Middle East 61 F3 27 20N 51 00E
Guwahati India 58 E3 26 10N 91 45E
GUYANA 84 D8
Gwadar Pakistan 58 B3 25 07N 62 19E
Gwalior India 58 C3 26 12N 78 09E
Gwynedd u.a. Wales 36 D2 52 45N 3 55W

H

Haarlem Netherlands 49 C5 52 23N 4 39E
Hachinohe Japan 59 N3 40 30N 141 30E
Haifa Israel 61 D4 32 49N 34 59E
Hail Saudi Arabia 61 E3 27 31N 41 45E
Hainan i. China 59 B1 18 50N 109 50E
Hai Phong Vietnam 60 B5 20 50N 106 41E
HAITI 79 M2
Hakodate Japan 59 N3 41 46N 140 44E
Halesowen England 37 E2 52 26N 2 05W
Halifax Canada 79 N5 44 38N 63 35W
Halifax England 34 D2 53 44N 1 52W
Halkirk Scotland 31 F3 58 30N 3 30W
Halle Germany 49 F4 51 28N 11 58E
Halmahera i. Indonesia 60 D3 0 30S 127 00E
Halton u.a. England 34 C2 53 20N 2 40W
Haltwhistle England 34 C3 54 58N 2 27W
Hamadan Iran 61 E4 34 46N 48 35E
Hamah Syria 43 P3 35 10N 36 45E
Hamamatsu Japan 59 M1 34 42N 137 42E
Hamburg Germany 49 F5 53 33N 10 00E
Hamersley Range Australia 72 B3 22 04S 117 37E
Hami China 58 E5 42 37N 93 32E
Hamilton Canada 79 M5 43 15N 79 50W
Hamilton New Zealand 73 D3 37 47S 175 17E
Hamilton Scotland 33 E3 55 47N 4 03W
Hamm Germany 49 D4 51 40N 7 49E
Hampshire co. England 38 C2 51 10N 1 15W
Hampshire Downs hills England 38 C2 51 15N 1 10W
Hangzhou China 59 D3 30 18N 120 07E
Hannover Germany 49 E5 52 23N 9 44E
Hanoi Vietnam 60 B5 21 01N 105 52E
Harad Saudi Arabia 61 E3 24 12N 49 12E
Harare Zimbabwe 67 D3 17 50S 31 03E
Harbin China 59 D4 45 45N 126 41E
Hargeysa Somalia 67 E5 9 31N 44 02E
Harlech Wales 36 C2 52 52N 4 07W
Harlow England 39 E2 51 47N 0 08E
Haroldswick Scotland 29 E3 60 47N 0 50W
Harrogate England 34 D2 54 00N 1 33W
Harrow England 38 D2 51 34N 0 20W
Hartland Point England 40 C4 51 02N 4 31W
Hartlepool England 35 D2 54 41N 1 13W
Hartlepool u.a. England 35 D3 54 41N 1 13W
Harwich England 39 F2 51 57N 1 17E
Harz mts. Europe 49 F4 52 00N 10 00E
Haslemere England 38 D1 51 06N 0 43W
Hassi Messaoud Algeria 66 C6 31 52N 5 43E
Hastings England 39 E1 50 51N 0 36E
Hastings New Zealand 73 D3 39 39S 176 51E
Hatherleigh England 40 C3 50 49N 4 04W
Hauraki Gulf New Zealand 73 D3 36 38S 175 04E
Havana Cuba 79 L3 23 07N 82 25W
Havant England 38 D1 50 51N 0 59W
Haverfordwest Wales 36 C1 51 49N 4 58W
Haverhill England 39 E2 52 05N 0 26E
Havering England 39 E2 51 34N 0 14E
Hawaiian Islands Pacific Ocean 4 20 00N 156 00W
Hawera New Zealand 73 C3 39 35S 174 17E
Hawick Scotland 33 G3 55 25N 2 47W
Hawke Bay New Zealand 73 D3 39 23S 177 12E
Hay-on-Wye Wales 36 D2 52 04N 3 07W
Hay River tn. Canada 78 H7 60 51N 115 42W
Haywards Heath England 38 D2 51 00N 0 06W
Hefei China 59 C3 31 55N 117 18E
Hegang China 59 E4 47 36N 130 30E
Heidelberg Germany 49 E3 49 25N 8 42E
Heilbronn Germany 49 E3 49 08N 9 14E
Hellín Spain 50 E3 38 31N 1 43W
Helmsdale Scotland 31 F3 58 07N 3 40W
Helmsdale r. Scotland 31 F3 58 10N 3 50W
Helsinki Finland 42 L8 60 08N 25 00E
Helston England 40 B3 50 05N 5 16W
Helvellyn mt. England 34 B3 54 32N 3 02W
Hemel Hempstead England 38 D2 51 46N 0 28W
Hemsworth England 35 D2 53 38N 1 21W
Hengelo Netherlands 49 D5 52 16N 6 46E
Henley-on-Thames England 38 D2 51 32N 0 56W
Herat Afghanistan 58 B4 34 20N 62 12E
Hereford England 37 E2 52 04N 2 43W
Herefordshire u.a. England 37 E2 52 12N 2 40W
Herma Ness c. Scotland 29 E3 60 50N 0 54W
Hermosillo Mexico 79 H3 29 15N 110 59W
Hertfordshire co. England 38 D2 51 50N 0 05W
Hexham England 34 C3 54 58N 2 06W
Heysham England 34 C3 54 02N 2 54W
Highland u.a. Scotland 30/31 D2/E2 57 00N 5 00W
High Veld mts. Republic of South Africa 67 C2 28 00S 28 00E
High Wycombe England 38 D2 51 38N 0 46W
Hilla Iraq 61 E4 32 28N 44 29E
Hillingdon England 38 D2 51 32N 0 27W
Hilversum Netherlands 49 C5 52 14N 5 10E
Himalaya mts. Asia 58 D4 28 00N 85 00E
Himeji Japan 59 L1 34 50N 134 40E
Hinckley England 37 F2 52 33N 1 21W
Hindu Kush mts. Afghanistan 58 B4 35 00N 70 00E
Hiroshima Japan 59 L1 34 23N 132 27E
Hitchin England 38 D2 51 57N 0 17W
Hobart Australia 72 E1 42 53S 147 19E
Hô Chi Minh Vietnam 60 B4 10 46N 106 43E
Hodeida Yemen Republic 61 E2 14 50N 42 58E
Höfn Iceland 42 C8 64 16N 15 10W
Hoggar mts. Algeria 66 C5 23 00N 6 00E
Hohhot China 59 C4 40 49N 111 37E
Hokkaido i. Japan 59 N3 43 30N 143 00E
Holderness p. England 35 E2 53 45N 0 05W
Holguín Cuba 79 M3 20 54N 76 15W
Hollywood USA 79 H4 34 05N 118 21W

Holsworthy England 40 C3 50 49N 4 21W
Holyhead Wales 36 C3 53 19N 4 38W
Holy Island England 33 H3 55 41N 1 48W
Holy Island Wales 36 C3 53 16N 4 39W
Homs Syria 61 D4 34 42N 36 40E
Homyel' Belarus 43 P3 52 25N 31 00E
HONDURAS 79 L2
Hong Kong China 59 C2 23 00N 114 00E
Honiara Solomon Islands 72 F5 9 28S 159 57E
Honiton England 41 D3 50 48N 3 13W
Honshu i. Japan 59 M2 37 15N 139 00E
Horn, Cape Chile 85 C2 56 00S 67 15W
Horncastle England 35 E2 53 13N 0 07W
Hornsea England 35 E2 53 55N 0 10W
Horsham England 38 D2 51 28N 0 21W
Hospitalet Spain 50 G4 41 21N 2 06E
Hounslow England 38 D2 51 28N 0 21W
Houston USA 79 K3 29 45N 95 25W
Hove England 38 D1 50 49N 0 11W
Howe, Cape Australia 72 E2 37 29S 149 59E
Howth Republic of Ireland 36 A3 53 23N 6 04W
Hoy i. Scotland 29 B1 58 48N 3 20W
Huainan China 59 C3 32 41N 117 06E
Huambo Angola 67 B3 12 44S 15 47E
Huancayo Peru 84 B6 12 05S 75 12W
Huang He r. China 59 C3 36 00N 116 00E
Hubli India 58 C2 15 20N 75 14E
Huddersfield England 34 D2 53 39N 1 47W
Hudson r. USA 79 M5 43 20N 73 50W
Hudson Bay Canada 78 L6/7 60 00N 89 00W
Hue Vietnam 60 B4 16 28N 107 35E
Huelva Spain 50 B2 37 15N 6 56W
Hufuf Saudi Arabia 61 E3 25 20N 49 34E
Hugh Town England 40 A2 49 55N 6 19W
Hull r. England 35 E2 53 50N 0 25W
Humber r. England 35 E2 53 40N 0 10W
HUNGARY 42 K5
Hungnam North Korea 59 D3 39 49N 127 40E
Hunstanton England 39 E3 52 57N 0 30E
Huntingdon England 39 E2 52 20N 0 12W
Huntly Scotland 31 G2 57 27N 2 47W
Huron, Lake Canada/USA 79 L5 45 00N 83 00W
Hyderabad India 58 C2 17 22N 78 26E
Hyderabad Pakistan 58 B3 25 22N 68 22E

I

Iaşi Romania 42 M5 47 09N 27 38E
Ibadan Nigeria 66 C3 7 23N 3 56E
Ibagué Colombia 84 B8 4 25N 75 20W
Ibiza Balearic Islands 50 F3 38 54N 1 26E
Ibiza i. Balearic Islands 50 F3 39 00N 1 20E
ICELAND 42 C9
Idaho state USA 79 H5 44 00N 115 00W
Ieper (Ypres) Belgium 39 G1 50 51N 2 53E
IJsselmeer l. Netherlands 49 C5 52 50N 5 15E
Ilchester England 41 E3 51 01N 2 41W
Ilebo Congo Dem. Rep. 67 C4 4 20S 20 35E
Ilfracombe England 40 C4 51 13N 4 08W
Ilhéus Brazil 84 F6 14 50S 39 06W
Ilkeston England 37 F2 52 59N 1 18W
Ilkley England 34 D2 53 56N 1 50W
Illinois state USA 79 L5 40 00N 89 00W
Ilminster England 41 E3 50 56N 2 55W
Iloilo The Philippines 60 D4 10 41N 122 33E
Immingham England 35 E2 53 36N 0 11W
Imphal India 58 E3 24 47N 93 55E
Inca Balearic Islands 50 G3 39 43N 2 54E
Inchon South Korea 59 D3 37 30N 126 38E
INDIA 58 C3
Indiana state USA 79 L5 40 00N 86 00W
Indianapolis USA 79 L4 39 45N 86 10W
Indian Ocean 7
INDONESIA 60 A3/D2
Indore India 58 C3 22 42N 75 54E
Indus r. Asia 58 B3 27 00N 68 00E
Ingleborough sum. England 34 C3 54 10N 2 23W
Ingolstadt Germany 49 F3 48 46N 11 27E
Inishowen Peninsula Republic of Ireland 32 B3 55 15N 7 15W
Inland Sea sd. Japan 59 L1 34 00N 132 30E
Inn r. Europe 49 G3 48 15N 12 30E
Inner Hebrides is. Scotland 30 C1/2 56 45N 6 45W
Innerleithen Scotland 33 F3 55 38N 3 05W
Inner Sound Scotland 30 D2 57 25N 5 55W
Innsbruck Austria 42 J5 47 17N 11 25E
Inuvik Canada 78 F7 68 16N 133 40W
Inveraray Scotland 32 D4 56 13N 5 05W
Inverbervie Scotland 31 G1 56 51N 2 17W
Invercargill New Zealand 73 B1 46 25S 168 22E
Inverclyde u.a. Scotland 33 E3 55 50N 4 45W
Invergarry Scotland 31 E2 57 02N 4 47W
Invergordon Scotland 31 E2 57 42N 4 10W
Inverkeithing Scotland 33 F4 56 02N 3 25W
Invermoriston Scotland 31 E2 57 13N 4 38W
Inverness Scotland 31 E2 57 27N 4 15W
Inverurie Scotland 31 G2 57 17N 2 23W
Iona i. Scotland 30 C1 56 19N 6 25W
Ionian Island Greece 42 L3 39 40N 20 51E
Ionian Sea Mediterranean Sea 42 K3 38 00N 27 00E
Iowa state USA 79 K5 42 00N 94 00W
Ipoh Malaysia 60 B3 4 36N 101 05E
Ipswich England 39 F3 52 04N 1 10E
Iquique Chile 84 B5 20 15S 70 08W
Iquitos Peru 84 B7 3 51S 73 13W
Iraklion Greece 42 M3 35 20N 25 08E
IRAN 61 F4
IRAQ 61 E4
Irbid Jordan 43 P2 32 33N 35 51E
Irian Jaya admin. Indonesia 72 D5 3 00S 133 00E
Irish Sea British Isles 36 D2 53 30N 5 30W
Irkutsk Russia 57 L3 52 18N 104 15E
Irrawaddy r. Myanmar 58 E3 26 00N 97 30E
Irthing r. England 34 C3 54 55N 2 40W
Irtysh r. Russia 56 H3 57 30N 72 30E
Irvine Scotland 33 E3 55 37N 4 40W
Ísafjördour Iceland 42 B9 66 05N 23 08W
Ishikari r. Japan 59 N3 43 30N 142 00E
Isigny-sur-Mer France 41 F2 49 18N 1 06W

Isla r. Scotland 31 F1 56 40N 3 15W
Islamabad Pakistan 58 C4 33 42N 73 10E
Island Magee p. Northern Ireland 32 D2 53 45N 5 45W
Islay i. Scotland 32 C3 55 48N 6 12W
Isle of Anglesey u.a. Wales 36 C3 53 18N 4 25W
Isle of Man British Isles 33 E2 54 15N 4 30W
Isle of Wight u.a. England 38 C1 50 40N 1 20W
ISRAEL 61 D4
Istanbul Turkey 42 M4 41 02N 28 57E
Istria p. Croatia 49 G1 45 00N 14 00E
Itabuna Brazil 84 F6 14 48S 39 18W
ITALY 51
Itchen r. England 38 C2 51 05N 1 15W
Ivanovo Russia 56 E3 57 00N 41 00E
Iwaki Japan 59 N2 37 03N 140 58E
Izhevsk Russia 43 V7 56 49N 53 11E
Izmir Turkey 42 M3 38 25N 27 10E
Izu Islands Japan 59 M1 34 20N 139 20E

J

Jabalpur India 58 C3 23 10N 79 59E
Jackson USA 79 K4 32 20N 90 11W
Jacksonville USA 79 L3 30 20N 81 40W
Jaffna Sri Lanka 58 D1 9 40N 80 01E
Jaipur India 58 C3 26 53N 75 50E
Jakarta Indonesia 60 B2 6 08S 106 45E
JAMAICA 79 M2
Jambi Indonesia 60 B2 1 34S 103 37E
James Bay Canada 78 L6 53 45N 81 00W
Jammu & Kashmir state Southern Asia 58 C4 29 40N 76 30E
Jamshedpur India 58 D3 22 47N 86 12E
JAPAN 59
Japan, Sea of Pacific Ocean 59 M3 39 00N 137 00E
Jauf Saudi Arabia 61 D3 29 49N 39 52E
Java i. Indonesia 60 B2/C2 7 00S 110 00E
Java Sea Indonesia 60 C2 5 00S 112 00E
Javalambre mt. Spain 50 E4 40 05N 1 02W
Jaya Peak Indonesia 72 D5 4 05S 137 09E
Jedburgh Scotland 33 G3 55 29N 2 34W
Jedda Saudi Arabia 61 D3 21 30N 39 10E
Jena Germany 49 F4 50 56N 11 35E
Jerez de la Frontera Spain 50 B2 36 41N 6 08W
Jersey i. Channel Islands British Isles 41 E2 49 13N 2 07W
Jerusalem Israel/Jordan 61 D4 31 47N 35 13E
Jiamusi China 59 E4 46 59N 130 29E
Jiayuguan China 59 A3 39 47N 98 14E
Jilin China 59 D4 43 53N 126 35E
Jinan China 59 C3 36 41N 117 00E
Jinsha Jiang (Yangtze) r. China 59 B2 27 30N 103 00E
Jixi China 59 E4 45 17N 131 00E
Jizan Saudi Arabia 61 E2 16 56N 42 33E
João Pessoa Brazil 84 F7 7 06S 34 53W
Johannesburg Republic of South Africa 67 C2 26 10S 28 02E
John o'Groats Scotland 31 F3 58 38N 3 05W
Johnstone Scotland 33 E3 55 50N 4 31W
Johor Bahru Malaysia 60 B3 1 27N 103 45E
Joinville Brazil 85 E5 26 20S 48 55W
Jönköping Sweden 42 J7 57 45N 14 10E
JORDAN 61 D4
Jos Plateau Nigeria 66 C3/4 9 30N 8 55E
Juan Fernández Islands Chile 85 B4 33 30S 80 00W
Juba Sudan 61 D1 4 50N 31 35E
Juba r. Somalia/Ethiopia 67 E5 0 00N 41 00E
Juiz de Fora Brazil 84 E5 21 47S 43 23W
Jungfrau mt. Switzerland 49 D2 46 33N 7 58E
Jura i. Scotland 32 D3/D4 55 55N 6 00W
Jura mts. France/Switzerland 49 F3 46 30N 6 00E
Jura, Sound of Scotland 32 D3/4 56 00N 5 45W
Jutland p. Denmark 49 E6 55 00N 9 00E

K

Kabul Afghanistan 58 B4 34 30N 70 00E
Kabwe Zambia 67 C3 14 29S 28 25E
Kaduna Nigeria 66 C4 10 28N 7 25E
Kaesong South Korea 59 D3 37 59N 126 30E
Kagoshima Japan 59 L1 31 37N 130 32E
Kainji Reservoir Nigeria 66 C4 10 25N 4 56E
Kaipara Harbour New Zealand 73 C3 36 12S 174 06E
Kaiserslautern Germany 49 D3 49 27N 7 47E
Kalahari Desert Southern Africa 67 C2 23 00S 23 00E
Kalemie Congo Dem. Rep. 67 C4 5 57S 29 10E
Kalgoorlie Australia 72 C2 30 49S 121 29E
Kaliningrad Russia 56 C3 54 40N 20 30E
Kamchatka Peninsula Russia 57 R3 57 30N 160 00E
Kampala Uganda 67 D5 0 19N 32 35E
Kananga Congo Dem. Rep. 67 C4 5 53S 22 26E
Kanazawa Japan 59 M2 36 35N 136 38E
Kandahar Afghanistan 58 B4 31 35N 65 45E
Kandalaksha Russia 42 N9 67 09N 32 31E
Kandla India 58 C3 23 03N 70 11E
Kandy Sri Lanka 58 D1 7 17N 80 40E
Kankan Guinea 66 B4 10 22N 9 11W
Kano Nigeria 66 C4 12 00N 8 31E
Kanpur India 58 D3 26 27N 80 14E
Kansas state USA 79 K4 38 00N 98 00W
Kansas City USA 79 K4 39 02N 94 33W
Kaohsiung Taiwan 59 D2 22 36N 120 17E
Kapuas r. Indonesia 60 C2 0 30N 112 00E
Karachi Pakistan 58 B3 24 52N 67 03E
Karaganda Kazakhstan 56 H2 49 53N 73 07E
Karbala Iraq 61 E4 32 37N 44 03E
Kariba, Lake Zambia/Zimbabwe 67 C3 17 00S 28 00E
Karlovac Croatia 51 E6 45 30N 15 34E
Karlsruhe Germany 49 E3 49 00N 8 24E
Karun r. Iran 61 E4 32 00N 48 00E
Kasai r. Angola/Congo Dem. Rep. 67 B4 4 00S 19 00E
Kashi China 58 C4 39 29N 76 02E
Kassala Sudan 61 D2 15 24N 36 30E
Kassel Germany 49 E4 51 18N 9 30E
Kasserine Tunisia 51 B1 35 13N 8 43E
Katherine Australia 72 D4 14 27S 132 15E

Kathmandu Nepal 58 D3 27 42N 85 19E
Katowice Poland 42 K6 50 15N 18 59E
Kaunas Lithuania 42 L6 54 52N 23 55E
Kayes Mali 66 A4 14 26N 11 28W
Kayseri Turkey 43 P3 38 42N 35 28E
KAZAKHSTAN 56 G2
Kazan' Russia 56 E3 55 45N 49 10E
Keady Northern Ireland 32 C2 54 15N 6 42W
Keighley England 34 D2 53 52N 1 54W
Keith Scotland 31 G2 57 32N 2 57W
Kelso Scotland 33 G3 55 36N 2 25W
Kemerovo Russia 56 J3 55 25N 86 05E
Kendal England 34 C3 54 20N 2 45W
Kenilworth England 37 F2 52 21N 1 34W
Kenitra Morocco 66 B6 34 20N 6 34W
Kennet r. England 38 C2 51 30N 1 45W
Kent co. England 39 E2 51 15N 0 09E
Kentucky state USA 79 L4 37 00N 85 00W
KENYA 67 D5
Kenya, Mount Kenya 67 D4 0 10S 37 19E
Kerema Papua New Guinea 72 E5 7 59S 145 46E
Kerguelen is. Indian Ocean 5 49 15N 69 10E
Kerman Iran 61 F4 30 18N 57 05E
Kermanshah Iran 61 E4 34 19N 47 04E
Kerrera i. Scotland 32 D1 56 24N 5 33W
Keswick England 34 B3 54 37N 3 08W
Kettering England 38 D2 52 24N 0 44W
Keynsham England 38 B2 51 26N 2 30W
Khabarovsk Russia 57 P2 48 32N 135 08E
Khaniá Greece 42 L3 35 31N 24 01E
Khanka, Lake Asia 57 P2 45 00N 132 30E
Kharkiv Ukraine 56 D2 50 00N 36 15E
Khartoum Sudan 61 D2 15 33N 32 35E
Kherson Ukraine 56 D2 46 39N 32 38E
Khulna Bangladesh 58 D2 22 49N 89 34E
Khyber Pass Afghanistan/Pakistan 58 C4 34 06N 71 05E
Kidderminster England 37 E2 52 23N 2 14W
Kidsgrove England 37 E3 53 05N 2 14W
Kidwelly Wales 36 C1 51 45N 4 18W
Kiel Germany 49 F6 54 20N 10 08E
Kielder Water res. England 33 G3 55 10N 2 30W
Kiev Ukraine 56 D3 50 25N 30 30E
Kigali Rwanda 67 D4 1 56S 30 04E
Kigoma Tanzania 67 C4 4 52S 29 36E
Kikwit Congo Dem. Rep. 67 B4 5 02S 18 51E
Kilbrannan Sound Scotland 32 D3 55 30N 5 35W
Kilkeel Northern Ireland 32 C2 54 04N 6 00W
Kilmacrenan Republic of Ireland 32 B3 55 02N 7 47W
Kilmaluag Scotland 30 C2 57 40N 6 17W
Kilmarnock Scotland 33 E3 55 36N 4 30W
Kilmory Scotland 32 D3 55 55N 5 41W
Kimberley Republic of South Africa 67 C2 28 45S 24 46E
Kinabalu, Mount Malaysia 60 C3 6 03N 116 32E
Kinbrace Scotland 31 F3 58 15N 3 56W
Kingsbridge England 40 D3 50 17N 3 46W
King's Lynn England 39 E3 52 45N 0 24E
Kingston Jamaica 79 M2 17 58N 76 48W
Kingston upon Hull England 35 E2 53 45N 0 20W
Kingston upon Hull u.a. England 35 E2 53 45N 0 20W
Kingston upon Thames England 38 D2 51 25N 0 18W
Kingswood England 38 B2 51 28N 2 30W
Kington England 36 D2 52 12N 3 01W
Kingussie Scotland 31 E2 57 05N 4 03W
Kinloch Scotland 30 C2 57 01N 6 17W
Kinross Scotland 31 F1 56 13N 3 27W
Kinshasa Congo Dem. Rep. 67 B4 4 18S 15 18E
Kintyre p. Scotland 32 D3 55 30N 5 35W
KIRIBATI 4/5
Kirkbean Scotland 33 F2 54 55N 3 36W
Kirkby England 34 C2 53 29N 2 54W
Kirkby Stephen England 34 C3 54 28N 2 20W
Kirkcaldy Scotland 33 F4 56 07N 3 10W
Kirkcudbright Scotland 33 E2 54 50N 4 03W
Kirkintilloch Scotland 33 E3 55 57N 4 10W
Kirk Michael Isle of Man British Isles 33 E2 54 17N 4 35W
Kirkpatrick, Mount Antarctica 86 84 25S 165 30E
Kirkuk Iraq 61 E4 35 28N 44 26E
Kirkwall Scotland 29 C1 58 59N 2 58W
Kirov Russia 56 E3 58 00N 49 38E
Kirriemuir Scotland 31 F1 56 41N 3 01W
Kisangani Congo Dem. Rep. 67 C5 0 33N 25 14E
Kisumu Kenya 67 D4 0 08S 34 47E
Kitakami r. Japan 59 N2 39 20N 141 20E
Kita Kyushu Japan 59 L1 33 52N 130 49E
Kitwe Zambia 67 C3 0 08S 30 30E
Kitzbühel Austria 51 D7 47 27N 12 23E
Kivu, Lake Congo Dem. Rep./Rwanda 67 C4 2 00S 29 00E
Klagenfurt Austria 49 H2 46 38N 14 20E
Klaipėda Lithuania 42 L7 55 43N 21 07E
Knaresborough England 35 D3 54 00N 1 27W
Knighton England 36 D2 52 21N 3 03W
Kobe Japan 59 M1 34 40N 135 12E
Koblenz Germany 49 D4 50 21N 7 36E
Kochi Japan 59 L1 33 33N 133 32E
Kodiak Island USA 78 D6 57 20N 153 40W
Kokand Uzbekistan 56 H2 40 33N 70 55E
Kola Peninsula Russia 42 P9 67 00N 39 00E
Kolhapur India 58 C2 16 40N 74 20E
Kolkata India 58 D2 22 30N 88 20E
Kolyma r. Russia 57 R4 68 00N 152 00E
Komsomolsk-on-Amur Russia 57 P3 50 32N 136 59E
Konya Turkey 43 O3 37 51N 32 30E
Koriyama Japan 59 N2 37 23N 140 22E
Kosciusko, Mount Australia 72 E2 36 27S 148 17E
Kota Bharu Malaysia 60 B3 6 08N 102 14E
Kota Kinabalu Malaysia 60 C3 5 59N 116 04E
Kotlas Russia 43 R8 61 15N 46 28E
Kragujevac Serbia and Montenegro 42 L4 44 01N 20 55E
Kra, Isthmus of Myanmar/Thailand 60 A4 10 20N 99 00E
Kraków Poland 42 K6 50 03N 19 55E
Kranj Slovenia 51 E7 46 15N 14 20E
Krasnodar Russia 56 D2 45 02N 39 00E
Krasnovodsk Turkmenistan 56 F2 40 01N 53 00E

Krasnoyarsk Russia 57 K3 56 05N 92 46E
Kratie Cambodia 60 B4 12 30N 106 03E
Krefeld Germany 49 D4 51 20N 6 32E
Krishna r. India 58 C2 16 00N 79 00E
Kryvyy Rih Ukraine 56 D2 47 55N 33 24E
Kuala Lumpur Malaysia 60 B3 3 09N 101 42E
Kuala Terengganu Malaysia 60 B3 5 20N 103 09E
Kuantan Malaysia 60 B3 3 48N 103 19E
Kuching Malaysia 60 C3 1 35N 110 21E
Kuh-e-Hazaran mt. Iran 61 F3 29 35N 57 20E
Kumamoto Japan 59 L1 32 50N 130 42E
Kumasi Ghana 66 B3 6 45N 1 35W
Kunlun Shan mts. China 58 D4 36 30N 85 00E
Kunming China 59 B2 25 04N 102 41E
Kupang Indonesia 60 D1 10 13S 123 38E
Kurgan Russia 56 G3 55 30N 65 20E
Kuril Islands Russia 57 R2 50 00N 155 00E
Kursk Russia 56 D3 51 45N 36 14E
Kushiro Japan 59 N3 42 58N 144 24E
Kutaisi Georgia 43 Q4 42 15N 42 44E
KUWAIT 61 E3
Kuwait Kuwait 61 E3 29 20N 48 00E
Kwangju South Korea 59 D3 35 07N 126 52E
Kyle of Lochalsh Scotland 30 D2 57 17N 5 43W
Kyoga, Lake Uganda 67 D5 2 00N 34 00E
Kyoto Japan 59 M2 35 02N 135 45E
KYRGYZSTAN 56 H2
Kyushu i. Japan 59 L1 32 20N 131 00E

L
Laâyoune Western Sahara 66 A5 27 10N 13 11W
Ladoga, Lake Russia 56 D4 61 00N 30 00E
Lae Papua New Guinea 72 E5 6 45S 147 00E
Lagos Nigeria 66 C3 6 27N 3 28E
La Grande Rivière r. Canada 78 L6 53 34N 74 36W
Lahore Pakistan 58 C4 31 35N 74 18E
Lairg Scotland 31 E3 58 01N 4 25W
Lake District geog. reg. England 34 B3/C3 54 30N 3 00W
Lakshadweep (Laccadive Islands) India 58 C2 9 30N 73 00E
Lambert Glacier Antarctica 86 73 00S 70 00E
Lambourn r. England 38 C2 51 30N 1 30W
Lammermuir Hills Scotland 33 G3 55 50N 2 45W
Lampedusa i. Italy 51 D1 35 00N 12 00E
Lampeter Wales 36 C2 52 07N 4 05W
Lanark Scotland 33 F3 55 41N 3 48W
Lancang Jiang (Mekong) r. China/Myanmar/ Laos/Cambodia 59 A3 32 00N 98 00E
Lancashire co. England 34 C2 53 50N 2 30W
Lancaster England 34 C3 54 03N 2 48W
Land's End c. England 40 B3 50 03N 5 44W
Languedoc geog. reg. France 48 E1/F2 44 00N 4 00E
Langholm Scotland 33 G3 55 09N 3 00W
Lanzhou China 59 B3 36 01N 103 45E
LAOS 60 B4
La Paz Bolivia 84 C6 16 30S 68 10W
La Plata Argentina 85 D4 34 52S 57 55W
La Puebla Balearic Islands 50 G3 39 46N 3 01E
Largs Scotland 32 E3 55 48N 4 52W
Lárisa Greece 42 L3 39 38N 22 25E
Larne Northern Ireland 32 D2 54 51N 5 49W
Larne district Northern Ireland 32 D2 54 50N 5 50W
la Rochelle France 48 C3 46 10N 1 10W
Larsen Ice Shelf Antarctica 86 67 00S 62 00W
la Sagra mt. Spain 50 D2 37 57N 2 34W
Las Palmas Canary Islands 66 A5 28 08N 15 27W
La Spezia Italy 51 B6 44 07N 9 48E
Las Vegas USA 79 H4 36 10N 115 10W
Latakia Syria 61 D4 35 31N 35 47E
Latina Italy 51 D4 41 28N 12 53E
LATVIA 42 M7
Launceston Australia 72 E1 41 27S 147 09E
Launceston England 40 C3 50 38N 4 21W
Laurencekirk Scotland 31 G1 56 50N 2 29W
Lausanne Switzerland 49 D2 46 32N 6 39E
Laval France 48 C4 48 04N 0 45W
Leader Water r. Scotland 33 G3 55 45N 2 50W
LEBANON 61 D4
Ledbury England 37 E2 52 02N 2 25W
Leeds England 34 D2 53 50N 1 35W
Leeuwin, Cape Australia 72 B2 34 21S 115 08E
le Havre France 48 D4 49 30N 0 06E
Leicester England 37 F2 52 38N 1 05W
Leicester City u.a. England 37 F2 52 36N 1 08W
Leicestershire co. England 37 F2/G2 52 30N 1 00W
Leiden Netherlands 49 C5 52 10N 4 30E
Leighton Buzzard England 38 D2 51 55N 0 41W
Leipzig Germany 49 G4 51 20N 12 25E
Leith Hill England 38 D2 51 11N 0 23W
Leitrim co. Republic of Ireland 32 A2 54 05N 8 00W
le Mans France 48 D4 48 00N 0 12E
Lena r. Russia 57 N4 64 30N 127 00E
Lens France 48 E5 50 26N 2 50E
Leominster England 37 E2 52 14N 2 45W
León Mexico 79 J3 21 10N 101 42W
León Spain 50 C5 42 35N 5 34W
Lerwick Scotland 29 D3 60 09N 1 09W
LESOTHO 67 C2
Lessay France 41 F2 49 13N 1 32W
Lesser Antilles is. West Indies 79 N2 18 00N 65 00W
Letchworth England 38 D2 51 58N 0 14W
le Touquet-Paris-Plage France 39 F1 50 31N 1 35E
Letterkenny Republic of Ireland 32 B2 54 57N 7 44W
Leverkusen Germany 49 D4 51 02N 6 59E
Levin New Zealand 73 D2 40 37S 175 17E
Lewes England 38 E1 50 52N 0 01E
Lewis i. Scotland 30 C3 58 15N 6 30W
Leyburn England 34 D3 54 19N 1 49W
Leyland England 34 C2 53 42N 2 42W
Leyte i. The Philippines 60 D4 11 00N 125 00E
Lhasa China 58 E3 29 40N 91 07E
Lianyungang China 59 C3 34 37N 119 10E
Liao He r. China 59 D4 42 30N 122 00E
Liard r. Canada 78 G7 61 00N 122 00W
LIBERIA 66 B3
Libreville Gabon 66 C3 0 30N 9 25E

LIBYA 66 D5
Libyan Desert North Africa 61 C3 25 00N 25 00E
Lichfield England 37 F2 52 42N 1 48W
Liddel Water r. England/Scotland 33 G3 55 15N 2 50W
LIECHTENSTEIN 49 E2
Liège Belgium 49 C4 50 38N 5 35E
Liepāja Latvia 42 L7 56 30N 21 00E
Likasi Congo Dem. Rep. 67 C3 10 58S 26 47E
Lille France 48 E5 50 39N 3 05E
Lillers France 39 G1 50 34N 2 29E
Lilongwe Malawi 67 D3 13 58S 33 49E
Lima Peru 84 B6 12 06S 8 40W
Limassol Cyprus 43 N2 34 04N 33 03E
Limavady Northern Ireland 32 C2 55 03N 6 57W
Limavady district Northern Ireland 32 B2/C2 54 50N 7 00W
Limerick Republic of Ireland 32 B2 52 40N 8 37W
Limoges France 48 D2 45 50N 1 15E
Limpopo r. Southern Africa 67 C2 22 30S 28 00E
Lincoln England 35 E2 53 14N 0 33W
Lincolnshire co. England 35 E2 53 10N 0 20W
Lincoln Wolds hills England 35 E2 53 25N 0 05W
Lingga i. Indonesia 60 B2 0 10S 104 40E
Linköping Sweden 42 K7 58 25N 15 35E
Linlithgow Scotland 33 F3 55 59N 3 37W
Linz Austria 42 J5 48 19N 14 18E
Lipetsk Russia 56 D3 52 37N 39 36E
Lisbon Portugal 50 A3 38 44N 9 08W
Lisburn Northern Ireland 32 C2 54 31N 6 03W
Lisburn district Northern Ireland 32 C2 54 30N 6 10W
Lisieux France 48 D4 49 09N 0 14E
Liskeard England 40 C3 50 28N 4 28W
Lismore i. Scotland 30 D1 56 30N 5 30W
LITHUANIA 42 L7
Littlehampton England 38 D1 50 48N 0 33W
Little Minch sd. Scotland 30 C2 57 45N 6 30W
Little Ouse r. England 39 E3 52 30N 0 30E
Little Rock USA 79 K4 34 42N 92 17W
Liuzhou China 59 B2 24 17N 109 15E
Liverpool England 34 C2 53 25N 2 55W
Livingston Scotland 33 F3 55 55N 3 32W
Livorno Italy 51 C5 43 33N 10 18E
Lizard England 40 B2 49 57N 5 13W
Lizard Point England 40 B2 49 56N 5 13W
Ljubljana Slovenia 51 E7 46 04N 14 30E
Llandeilo Wales 36 D1 51 53N 3 59W
Llandovery Wales 36 D1 51 59N 3 48W
Llandrindod Wells Wales 36 D2 52 15N 3 23W
Llandudno Wales 36 D3 53 19N 3 49W
Llanelli Wales 36 C1 51 42N 4 10W
Llangollen Wales 36 D2 52 58N 3 10W
Llanidloes Wales 36 D2 52 27N 3 32W
Lleida Spain 50 F4 41 37N 0 38E
Lleyn Peninsula Wales 36 C2 52 53N 4 30W
Lobito Angola 67 B3 12 20S 13 34E
Lochaline tn. Scotland 30 D1 56 32N 5 47W
Loch Arkaig l. Scotland 30 D1 56 59N 5 10W
Loch Awe l. Scotland 30/31 D1 56 15N 5 17W
Lochdon tn. Scotland 30 D1 56 25N 5 35W
Loch Doon l. Scotland 33 E3 55 26N 4 38W
Loch Earn l. Scotland 31 E1 56 25N 4 10W
Loch Eishort est. Scotland 30 D2 57 08N 6 00W
Loch Eriboll b. Scotland 31 E3 58 31N 4 41W
Loch Ericht l. Scotland 31 E1 56 48N 4 25W
Loch Etive l. Scotland 30/31 D1 56 30N 5 10W
Loch Fyne b. Scotland 32 D4 56 10N 5 05W
Lochgilphead tn. Scotland 32 D2 56 03N 5 26W
Loch Katrine l. Scotland 33 E4 56 16N 4 30W
Loch Ken l. Scotland 33 E3 55 02N 4 07W
Loch Leven l. Scotland 31 F1 56 15N 3 23W
Loch Linnhe b. Scotland 30 D1 56 35N 5 25W
Loch Lochy l. Scotland 31 E1 56 58N 4 55W
Loch Lomond l. Scotland 33 E4 56 10N 4 35W
Loch Long l. Scotland 32/33 E4 56 10N 4 50W
Lochmaben tn. Scotland 33 F3 55 08N 3 27W
Lochmaddy tn. Scotland 30 B2 57 36N 7 08W
Loch Nan Clar l. Scotland 31 E3 58 30N 4 00W
Loch Ness l. Scotland 31 E2 57 20N 4 30W
Loch Rannoch l. Scotland 31 E1 56 41N 4 18W
Loch Shiel l. Scotland 30 D1 56 45N 5 35W
Loch Shin l. Scotland 31 E3 58 05N 4 30W
Loch Snizort b. Scotland 30 C2 57 30N 6 30W
Loch Tay l. Scotland 31 E1 56 31N 4 10W
Loch Torridon b. Scotland 30 D2 57 35N 5 46W
Lockerbie Scotland 33 F3 55 07N 3 22W
Logan, Mount Canada 78 E6 60 34N 140 25W
Logroño Spain 50 D5 42 28N 2 26W
Loire r. France 48 D3 47 30N 1 00E
Lolland i. Denmark 49 F6 54 45N 12 20E
Lombok i. Indonesia 60 C2 8 30S 116 30E
Lomé Togo 66 C3 6 10N 1 21E
London England 38 D2 51 30N 0 10W
Londonderry Northern Ireland 32 B2 54 59N 7 19W
Londonderry district Northern Ireland 32 B2 54 55N 7 20W
Londrina Brazil 84 D5 23 18S 51 13W
Long Eaton England 37 F2 52 54N 1 15W
Long Island The Bahamas 79 M3 23 20N 75 00W
Longreach Australia 72 E3 23 26S 144 15E
Longs Peak USA 79 J5 40 20N 105 50W
Looe England 40 C3 51 21N 4 27W
Lorient France 48 B3 47 45N 3 21W
Los Angeles USA 79 H4 34 00N 118 15W
Lossiemouth Scotland 31 F2 57 43N 3 18W
Lostwithiel England 40 C3 50 25N 4 40W
Lot r. France 48 E2 44 30N 2 15E
Louangphrabang Laos 60 B4 19 53N 102 10E
Lough Allen l. Republic of Ireland 32 A2 54 15N 8 00W
Lough Beg l. Northern Ireland 32 C2 54 48N 6 16W
Loughborough England 37 F2 52 47N 1 11W
Lough Derg l. Republic of Ireland 32 B2 52 55N 8 15W

Lough Foyle est. Northern Ireland/Republic of Ireland 32 B3 55 10N 7 10W
Lough Macnean Lower l. Northern Ireland 32 B2 54 17N 7 50W
Lough Macnean Upper l. Northern Ireland/ Republic of Ireland 32 B2 54 18N 7 55W
Lough Melvin l. Northern Ireland/Republic of Ireland 32 A2 54 25N 8 05W
Lough Neagh l. Northern Ireland 32 C2 54 35N 6 30W
Lough Oughter l. Republic of Ireland 32 B2 54 00N 7 30W
Lough Swilly b. Republic of Ireland 32 B3 55 20N 7 35W
Louisiana state USA 79 K4 32 00N 92 00W
Louisville USA 79 L4 38 13N 85 48W
Lourdes France 48 C1 43 06N 0 02W
Louth England 35 E2 53 22N 0 01W
Louth co. Republic of Ireland 32 C1/2 53 55N 6 25W
Lower California p. Mexico 79 H3 27 30N 113 00W
Lower Lough Erne l. Northern Ireland 32 B2 54 25N 7 45W
Lowestoft England 39 F3 52 29N 1 45E
Lowther Hills Scotland 33 F3 55 25N 3 45W
Lualaba r. Congo Dem. Rep 67 C4 4 00S 26 30E
Luanda Angola 67 B4 8 50S 13 15E
Lubango Angola 67 B3 14 55S 13 30E
Lübeck Germany 49 F5 53 52N 10 40E
Lublin Poland 42 L6 51 18N 22 31E
Lubumbashi Congo Dem. Rep 67 C3 11 41S 27 29E
Luce Bay Scotland 32/33 E2 54 47N 4 50W
Lucerne Switzerland 49 E2 47 03N 8 17E
Lucknow India 58 D3 26 50N 80 54E
Lüderitz Namibia 67 B2 26 38S 15 10E
Ludhiana India 58 C4 30 56N 75 52E
Ludlow England 37 E2 52 22N 2 43W
Ludwigshafen Germany 49 E3 49 29N 8 27E
Lugano Switzerland 49 E2 46 01N 8 57E
Luhans'k Ukraine 56 E2 48 35N 39 20E
Luleå Sweden 42 L9 65 35N 22 10E
Lundy i. England 40 C4 51 11N 4 40W
Lune r. England 34 C3 54 15N 2 40W
Luoyang China 59 C3 34 47N 112 26E
Lurgan Northern Ireland 32 C2 54 28N 6 20W
Lusaka Zambia 67 C3 15 26S 28 20E
Luton England 38 D2 51 53N 0 25W
Luton u.a. England 38 D2 51 53N 0 30W
LUXEMBOURG 49 C3/D3
Luxembourg Luxembourg 49 D3 49 37N 6 08E
Luxor Egypt 61 D3 25 41N 32 24E
Luzon i. The Philippines 60 D4 15 00N 122 00E
L'viv Ukraine 56 C2 49 50N 24 00E
Lybster Scotland 31 F3 58 18N 3 13W
Lyme Bay England 41 E3 50 40N 2 55W
Lyme Regis England 41 E3 50 44N 2 57W
Lymington England 38 C1 50 46N 1 33W
Lynn Lake tn. Canada 78 J6 56 51N 101 01W
Lynton England 40 D4 51 15N 3 50W
Lyons France 48 F2 45 46N 4 50E
Lys r. France 39 G1 50 40N 2 10E
Lytham St. Anne's England 34 B2 53 45N 3 01W

M
Maas r. Netherlands 49 C4 51 50N 5 45E
Maastricht Netherlands 49 C4 50 51N 5 42E
Mablethorpe England 35 F2 53 21N 0 15E
Macao China 59 C2 22 10N 113 35E
Macapá Brazil 84 D8 0 04N 51 04W
Macclesfield England 34 C2 53 16N 2 07W
Macdonnell Ranges Australia 72 D3 23 42S 128 42E
Macduff Scotland 31 G2 57 40N 2 29W
MACEDONIA, FORMER YUGOSLAV REPUBLIC OF (FYROM) 42 L4
Maceió Brazil 84 F7 9 40S 35 44W
Machynlleth Wales 36 D2 52 35N 3 51W
Mackay, Lake Australia 72 C3 22 25S 128 42E
Mackenzie r. Canada 78 G7 64 00N 124 00W
Mackenzie Mountains Canada 78 G7 66 00N 132 00W
McKinley, Mount USA 78 E7 62 02N 151 01W
Mâcon France 48 F3 46 18N 4 50E
MADAGASCAR 67 E2/3
Madang Papua New Guinea 72 E5 5 14S 145 45E
Madeira r. Brazil 84 C7 6 00S 61 30W
Madeira Islands Atlantic Ocean 5 32 45N 17 00W
Madrid Spain 50 D4 40 25N 3 43W
Madura i. Indonesia 60 C2 7 00S 113 00E
Madurai India 58 C1 9 55N 78 07E
Maebashi Japan 59 M2 36 24N 139 04E
Magadan Russia 57 R3 59 38N 150 50E
Magdalena r. Colombia 84 B8 3 00N 75 40W
Magdeburg Germany 49 F5 52 08N 11 37E
Maggiore, Lake Italy 51 B6/7 46 00N 8 00E
Maghera Northern Ireland 32 C2 54 51N 6 40W
Magherafelt Northern Ireland 32 C2 54 45N 6 36W
Magherafelt district Northern Ireland 32 C2 54 45N 6 50W
Magnitogorsk Russia 56 F3 53 28N 59 06E
Mahajanga Madagascar 67 E3 15 40S 46 20E
Mahakam r. Indonesia 60 C3 0 30N 116 00E
Mahilyow Belarus 56 D3 53 54N 30 20E
Maidenhead England 38 D2 51 32N 0 44W
Maidstone England 39 E2 51 17N 0 32E
Maiduguri Nigeria 66 D3 11 53N 13 16E
Main r. Germany 49 F4 50 00N 10 30E
Mai-Ndombe, Lake Congo Dem. Rep. 67 B4 2 00S 18 20E
Maine state USA 79 N5 45 00N 70 00W
Mainland i. Orkney Islands Scotland 29 B1/B2 59 00N 3 15W
Mainland i. Shetland Islands Scotland 29 D3 60 15N 1 20W
Mainz Germany 49 D3 50 00N 8 16E
Makassar Strait Indonesia 60 C2 0 00 119 00E
Makeyevka Ukraine 56 D2 48 01N 38 00E
Makhachkala Russia 56 E2 42 59N 47 30E
Malabo Equatorial Guinea 66 C3 3 45N 8 48E
Malacca, Strait of Indonesia 60 B3 4 00N 100 00E

Málaga Spain 50 C2 36 43N 4 25W
Malahide Republic of Ireland 36 A3 53 27N 6 09W
Malakal Sudan 61 D1 9 31N 31 40E
Malang Indonesia 60 C2 7 59S 112 45E
Malanje Angola 67 B4 9 32S 16 20E
Malatya Turkey 43 P3 38 22N 38 18E
MALAWI 67 D3
Malawi, Lake see Nyasa, Lake
Malaya admin. Malaysia 60 B3 5 00N 102 00E
MALAYSIA 60 B3/C3
MALDIVES 58 C1
MALI 66 B4
Malin Head c. Republic of Ireland 32 B3 55 30N 7 20W
Mallaig Scotland 30 D2 57 00N 5 50W
Mallorca i. Balearic Islands 50 G3 39 50N 2 30E
Malmesbury England 38 B2 51 36N 2 06W
Malmö Sweden 42 J7 55 35N 13 00E
MALTA 51 E1
Malton England 35 E3 54 08N 0 48W
Manacor Balearic Islands 50 G3 39 35N 3 12E
Manado Indonesia 60 D3 1 32N 124 55E
Managua Nicaragua 79 L2 12 06N 86 18W
Manaus Brazil 84 D7 3 06S 60 00W
Manawatu r. New Zealand 73 D2 40 30S 176 00E
Manchester England 34 C2 53 30N 2 15W
Mandalay Myanmar 58 E3 21 57N 96 04E
Manfredonia Italy 51 E4 41 37N 15 55E
Mangalore India 58 C2 12 54N 74 51E
Mangotsfield England 38 B2 51 29N 2 31W
Manila The Philippines 60 D4 14 37N 120 58E
Manitoba province Canada 78 K6 55 00N 98 00W
Manizales Colombia 84 B8 5 03N 75 32W
Mannheim Germany 49 E3 49 30N 8 28E
Mansfield England 37 F3 53 09N 1 11W
Manzanares Spain 50 D3 39 00N 3 23W
Maputo Mozambique 67 D2 25 58S 32 35E
Maracaibo Venezuela 84 B9 10 44N 71 37W
Maracaibo, Lake Venezuela 84 B9 9 50N 71 30W
Marañon r. Peru 84 B7 7 00S 77 40W
Marbella Spain 50 C2 36 31N 4 53W
Marble Bar Australia 72 B3 21 10S 119 43E
Mar del Plata Argentina 85 D4 38 00S 57 32W
Marden England 39 E2 51 11N 0 30E
Margarita i. Venezuela 79 N2 11 30N 64 00W
Margate England 39 F2 51 24N 1 24E
Maribor Slovenia 51 E7 46 34N 15 38E
Marie Byrd Land geog. reg. Antarctica 86 77 00S 130 00W
Mariscal Estigarribia Paraguay 84 C5 22 03S 60 35W
Mariupol Ukraine 56 D2 47 05N 37 34E
Market Drayton England 37 E2 52 54N 2 29W
Market Harborough England 37 G2 52 29N 0 55W
Market Rasen England 35 E2 53 24N 0 21W
Markham, Mount Antarctica 86 82 45S 160 25E
Marlborough England 38 C2 51 26N 1 43W
Marlow England 38 D2 51 35N 0 48W
Marne r. France 48 F4 48 50N 3 45E
Marrakesh Morocco 66 B6 31 49N 8 00W
Marsala Italy 51 D2 37 48N 12 27E
Marseille France 48 F1 43 18N 5 22E
MARSHALL ISLANDS 5
Martinique i. Lesser Antilles 79 N2 14 30N 61 00W
Maryborough Australia 72 F3 25 33S 152 42E
Maryland state USA 79 M4 39 00N 77 00W
Maryport England 34 B3 54 43N 3 30W
Maseru Lesotho 67 C2 29 19S 27 29E
Mashhad Iran 61 F4 36 16N 59 34E
Massachusetts state USA 79 M5 42 00N 72 00W
Massawa Eritrea 61 D2 15 42N 39 25E
Massif Central mts. France 48 E2 45 00N 3 30E
Masterton New Zealand 73 D2 40 57S 175 39E
Matadi Congo Dem. Rep. 67 B4 5 50S 13 32E
Matamoros Mexico 79 K3 25 33N 103 51W
Matlock England 37 F3 53 08N 1 32W
Mato Grosso geog. reg. Brazil 84 D6 14 00S 56 00W
Matsue Japan 59 L2 35 29N 133 04E
Matsuyama Japan 59 L1 33 50N 132 47E
Matterhorn mt. Switzerland 49 D1 45 59N 7 39E
MAURITANIA 66 A5
MAURITIUS 5
Maybole Scotland 33 E3 55 21N 4 41W
Mazatlán Mexico 79 J3 23 11N 106 25W
Mbandaka Congo Dem. Rep. 67 B5 0 03N 18 28E
Mbuji-Mayi Congo Dem. Rep. 67 C4 6 10S 23 39E
Mecca Saudi Arabia 61 D3 21 26N 39 49E
Medan Indonesia 60 A3 3 35N 98 39E
Medellín Colombia 84 B8 6 15N 75 36W
Medina Saudi Arabia 61 D3 24 30N 39 35E
Mediterranean Sea Africa/Europe 42/43 35 00N 15 00E
Medway r. England 39 E2 51 24N 0 30E
Medway u.a. England 39 E2 51 24N 0 30E
Meekatharra Australia 72 B3 26 35S 118 30E
Meerut India 58 C3 29 00N 77 42E
Mega Ethiopia 61 D1 4 02N 38 19E
Meharry, Mount Australia 72 B3 22 58S 118 35E
Meig r. Scotland 31 E2 57 40N 4 50W
Meknes Morocco 66 B6 33 53N 5 37W
Mekong (Lancang Jiang) r. Cambodia/Laos/ Myanmar/China 59 B1 15 00N 105 00E
Melbourne Australia 72 E2 37 49S 144 58E
Melrose Scotland 33 G3 55 36N 2 44W
Melton Mowbray England 37 G2 52 46N 0 53W
Melville Island Canada 78 H9 75 30N 112 00W
Memphis USA 79 K4 35 10N 90 00W
Menam r. Thailand 58 B4 15 00N 100 00E
Mendip Hills England 41 E4 51 18N 2 45W
Mendoza Argentina 85 C4 32 48S 68 52W
Menorca i. Balearic Islands 50 H3 39 45N 4 15E
Mentawai Islands Indonesia 60 A2 2 00N 99 00E
Menzies, Mount Antarctica 86 73 30S 62 00E
Meon r. England 38 C1 50 55N 1 10W
Mere England 38 B2 51 06N 2 16W
Mérida Mexico 79 L3 20 59N 89 39W
Mersey r. England 34 C2 53 20N 2 55W
Merseyside co. England 34 C2 53 30N 3 20W
Mersin Turkey 43 N3 36 47N 34 37E
Merthyr Tydfil Wales 36 D1 51 46N 3 23W
Merthyr Tydfil u.a. Wales 36 D1 51 48N 3 20W
Merton England 38 D2 51 25N 0 12W

Messina Italy **51** E3 38 13N 15 33E
Metz France **48** G4 49 07N 6 11E
Meuse r. Belgium/France **49** C4 50 30N 5 00E
Mexicali Mexico **79** H4 32 36N 115 30E
MEXICO 79 H4/K2
Mexico City Mexico **79** K2 19 25N 99 10W
Mexico, Gulf of Mexico **79** L3 25 00N 90 00W
Miami USA **79** L3 25 45N 80 15W
Michigan state USA **79** L5 45 00N 85 00W
Michigan, Lake Canada/USA **79** L5 45 00N 87 00W
Mickle Fell sum. England **34** C3 54 38N 2 18W
Middlesbrough England **35** D3 54 35N 1 14W
Middlesbrough u.a. England **35** D3 54 33N 1 04W
Midlothian u.a. Scotland **33** F3 55 50N 3 05W
Milan Italy **51** B6 45 28N 9 12E
Mildenhall England **39** E3 52 21N 0 30E
Mildura Australia **72** E2 34 12S 142 09E
Milford Haven Wales **36** B1 51 44N 5 02W
Milford Sound New Zealand **73** B2 44 41S 167 56E
Milton Keynes England **38** D3 52 02N 0 42W
Milton Keynes u.a. England **38** D3 52 10N 0 40W
Milton Ness c. Scotland **31** G1 56 46N 2 20W
Milwaukee USA **79** L5 43 03N 87 56W
Minch, The sd. Scotland **30** C3/D3 58 00N 6 00W
Mindanao i. The Philippines **60** D3 8 00N 125 00E
Mindoro i. The Philippines **60** D4 13 00N 121 00E
Minehead England **40** D4 51 13N 3 29W
Mingulay i. Scotland **30** B1 56 49N 7 38W
Minneapolis USA **79** K5 45 00N 93 15W
Minnesota state USA **79** K5 47 00N 95 00W
Minsk Belarus **56** C3 53 51N 27 30E
Miskolc Hungary **42** L5 48 07N 20 47E
Misool i. Indonesia **60** E2 1 50S 129 55E
Misratah Libya **66** D6 32 23N 15 06E
Mississippi r. USA **79** K4 33 00N 90 30W
Mississippi state USA **79** K3 32 00N 90 00W
Missouri r. USA **79** J5 48 00N 109 00W
Missouri state USA **79** K4 38 00N 93 00W
Miyazaki Japan **59** L1 31 56N 131 27E
Mobile USA **79** L4 30 40N 88 05W
Moçambique Mozambique **67** E3 15 03S 40 45E
Módena Italy **51** C6 44 39N 10 55E
Moffat Scotland **33** F3 55 20N 3 27W
Mogadishu Somalia **67** E5 2 02N 45 21E
Mold Wales **36** D3 53 10N 3 08W
Moldau r. Czech Republic **49** H3 49 30N 14 15E
MOLDOVA 42 M5
Moluccas is. Indonesia **60** D3/E2 4 00S 127 00E
Molucca Sea Indonesia **60** D2 0 30S 125 30E
Mombasa Kenya **67** D4 4 04S 39 40E
MONACO 48 G1
Monadhliath Mountains Scotland **31** E2 57 10N 4 00W
Monaghan Republic of Ireland **32** C2 54 15N 6 58W
Monaghan co. Republic of Ireland **32** B2/C2 54 10N 7 00W
Monforte Spain **50** B5 42 32N 7 30W
MONGOLIA 57 L2
Monmouth Wales **37** E1 51 50N 2 43W
Monmouthshire u.a. Wales **37** E1 51 47N 2 55W
Monopoli Italy **51** F4 40 57N 17 18E
Monrovia Liberia **66** A3 6 20N 10 46W
Montana state USA **79** H5 47 00N 111 00W
Montauban France **48** D2 44 01N 1 20E
Montbéliard France **48** G3 47 31N 6 48E
Mont Blanc mt. France/Italy **42** G5 45 50N 6 52E
Monte Carlo Monaco **48** G1 43 44N 7 25E
Monterrey Mexico **79** J3 25 40N 100 20W
Montes Claros tn. Brazil **84** E6 16 45S 43 52W
Montevideo Uruguay **85** D4 34 55S 56 10W
Montgomery Wales **36** D2 52 34N 3 10W
Montluçon France **48** E3 46 20N 2 36E
Montpellier France **48** E1 43 36N 3 53E
Montréal Canada **79** M5 45 32N 73 36W
Montreux Switzerland **42** H4 46 27N 6 55E
Montrose Scotland **31** G1 56 43N 2 29W
Monza Italy **51** B6 45 35N 9 16E
Moosonee Canada **79** L6 51 18N 80 39W
Moray u.a. Scotland **31** F2 57 20N 3 10W
Moray Firth est. Scotland **31** F2 57 45N 3 45W
Morecambe England **34** C3 54 04N 2 53W
Morecambe Bay England **34** B3/C3 54 00N 3 00W
Morelia Mexico **79** J2 19 40N 101 11W
Moreton-in-Marsh England **38** C2 51 59N 1 42W
Morioka Japan **59** N2 39 43N 141 08E
Morlaix France **48** B4 48 35N 3 50W
MOROCCO 66 B6
Moroni Comoros **67** E3 11 40S 43 16E
Morotai i. Indonesia **60** D3 2 30N 128 30E
Morven mt. Scotland **31** F3 58 14N 3 42W
Moscow Russia **56** D3 55 45N 37 42E
Mosel r. Germany/France **49** G3 50 00N 7 00E
Móstoles Spain **50** D4 40 19N 3 53W
Mosul Iraq **61** E4 36 21N 43 08E
Motherwell Scotland **33** E3 55 48N 3 59W
Moulmein Myanmar **58** E2 16 30N 97 39E
Mount Hagen tn. Papua New Guinea **72** E5 5 54S 114 13E
Mount Isa tn. Australia **72** D3 20 44S 139 29E
Mount Magnet tn. Australia **72** B3 28 02S 117 50E
Mounts Bay England **40** B3 50 05N 5 30W
Moyale Kenya **66** D5 3 31N 39 04E
Moyle district Northern Ireland **32** C3 55 10N 6 10W
MOZAMBIQUE 67 D3
Mozambique Channel Mozambique/Madagascar **67** E3 18 00S 42 00E
Muck i. Scotland **30** B2 56 50N 6 15W
Muckle Roe i. Scotland **29** D3 60 22N 1 26W
Mudanjiang China **59** D4 44 36N 129 42E
Mukalla Yemen Republic **61** E2 14 34N 49 09E
Mulhacén mt. Spain **50** D2 37 04N 3 19W
Mulhouse France **48** G3 47 45N 7 21E
Mull i. Scotland **30** C1/D1 56 25N 6 00W
Mull Head c. Scotland **29** C2 59 20N 2 58W
Mullion England **40** B3 50 01N 5 14W
Mull of Galloway c. Scotland **32** E2 54 38N 4 50W
Mull of Kintyre c. Scotland **32** D3 55 17N 5 55W
Mull of Oa c. Scotland **32** C3 55 35N 6 20W
Mull, Sound of Scotland **30** D1 56 30N 6 00W
Multan Pakistan **58** C4 30 10N 71 36E

Mumbai India **58** C2 18 56N 72 51E
München Gladbach Germany **49** D4 51 12N 6 25E
Munich Germany **49** F3 48 08N 11 35E
Münster Germany **49** D4 51 58N 7 37E
Mur r. Europe **49** H2 47 00N 14 00E
Murcia Spain **50** E2 37 59N 1 08W
Murmansk Russia **56** D4 68 59N 33 08E
Muroran Japan **59** N3 42 21N 140 59E
Murray r. Australia **72** E2 35 40S 144 00E
Murrumbidgee r. Australia **72** E2 34 00S 145 00E
Muscat Oman **61** F3 23 37N 58 38E
Musi r. Indonesia **60** B2 3 00S 104 00E
Musselburgh Scotland **33** F3 55 57N 3 04W
Mwanza Tanzania **67** D4 2 31S 32 56E
Mweru, Lake Congo Dem. Rep./Zambia **67** C4 8 30S 28 30E
MYANMAR (BURMA) **58** E3
Mykolayiv Ukraine **56** D2 46 57N 32 00E
Mynydd Eppynt hills Wales **36** D2 52 02N 3 35W
Mynydd Preseli hills Wales **36** C1 51 58N 4 45W
Mysore India **58** C2 12 18N 76 37E

N

Naberezhnyye Chelny Russia **56** F3 55 42N 52 19E
Nabeul Tunisia **42** J3 36 30N 10 44E
Nador Morocco **50** D1 35 10N 3 00W
Nagano Japan **59** M2 36 39N 138 10E
Nagasaki Japan **59** K1 32 45N 129 52E
Nagoya Japan **59** M2 35 08N 136 53E
Nagpur India **58** C3 21 10N 79 12E
Nairn Scotland **31** F2 57 35N 3 53W
Nairn r. Scotland **31** E2 57 28N 4 10W
Nairobi Kenya **67** D4 1 17S 36 50E
Najran Saudi Arabia **61** E2 17 37N 44 40E
Namangan Uzbekistan **56** H2 40 59N 71 41E
Namib Desert Namibia **67** B2 22 00S 14 00E
NAMIBIA 67 B2
Nampo North Korea **59** D3 38 51N 125 10E
Nampula Mozambique **66** D3 15 09S 39 14E
Namur Belgium **49** C4 50 28N 4 52E
Nanchang China **59** C2 28 33N 115 58E
Nancy France **48** G4 48 42N 6 12E
Nanjing China **59** C3 32 03N 118 47E
Nanning China **59** B2 22 50N 108 19E
Nan Shan mts. China **59** A3 38 00N 99 00E
Nantes France **48** C3 47 14N 1 35W
Napier New Zealand **73** D3 39 30S 176 54E
Naples Italy **51** E4 40 50N 14 15E
Napo r. Ecuador/Peru **84** B7 1 30S 74 30W
Narbonne France **48** E1 43 11N 3 00E
Narmada r. India **58** C3 22 00N 78 00E
Narvik Norway **42** K9 68 26N 17 25E
Nassau The Bahamas **79** M3 25 05N 77 20W
Nasser, Lake Egypt **61** D3 22 35N 31 40E
Natal Brazil **84** F7 5 46S 35 15W
Natuna Islands Indonesia **60** B3 3 40N 108 00E
NAURU 5
Ndjamena Chad **66** D4 12 10N 14 59E
Ndola Zambia **67** C3 13 00S 28 39E
Neath Wales **36** D1 51 40N 3 48W
Neath Port Talbot u.a. Wales **36** D1 51 40N 3 45W
Nebraska state USA **79** J5 42 00N 102 00W
Negro r. Argentina **85** C4 39 39S 67 30W
Negro r. Bolivia/Brazil **84** C7 1 00S 62 20W
Negros i. The Philippines **60** D3 10 00N 123 00E
Neiva Colombia **84** B8 2 58N 75 15W
Nelson England **34** C2 53 51N 2 13W
Nelson New Zealand **73** C2 41 30S 172 30E
Nelson r. Canada **78** K6 57 00N 94 00W
Nemuro Japan **59** P3 43 22N 145 36E
Nene r. England **39** E3 52 45N 0 10E
Nen Jiang r. China **59** D4 45 00N 124 00E
NEPAL 58 D3
NETHERLANDS 49 C5/D5
Nevada state USA **79** H4 39 00N 118 00W
Nevers France **48** E3 47 00N 3 09E
Nevis i. St. Kitts and Nevis **79** N2 17 00N 62 30W
Newark USA **79** M5 40 43N 74 11W
Newark-on-Trent England **37** G3 53 05N 0 49W
New Britain i. Papua New Guinea **72** F5 4 45S 150 30E
New Brunswick province Canada **79** N5 47 30N 66 00W
Newburgh Scotland **33** F4 57 18N 2 00W
Newbury England **38** C2 51 25N 1 20W
New Caledonia i. Pacific Ocean **5** G3 20 00S 165 00E
Newcastle Australia **72** F2 32 57S 151 43E
Newcastle Northern Ireland **32** D2 54 12N 5 54W
Newcastle Emlyn Wales **36** C2 52 02N 4 28W
Newcastle-under-Lyme England **37** E3 53 00N 2 14W
Newcastle upon Tyne England **34** D3 54 59N 1 35W
New Cumnock Scotland **33** E3 55 24N 4 12W
Newfoundland i. Canada **79** P5 53 51N 56 56W
Newfoundland and Labrador province Canada **78** P6 56 00N 60 00W
New Galloway Scotland **33** E3 55 05N 4 10W
New Hampshire state USA **79** M5 43 00N 72 00W
New Ireland i. Papua New Guinea **72** F5 3 15S 152 30E
New Jersey state USA **79** M4 40 00N 75 00W
Newman Australia **72** B3 23 22S 119 43E
Newmarket England **39** E2 52 15N 0 25E
New Mexico state USA **79** J4 35 00N 107 00W
New Orleans USA **79** K3 30 00N 90 03W
Newport Isle of Wight England **38** C1 50 42N 1 18W
Newport u.a. Wales **36** D1/E1 51 33N 3 00W
Newport Telford and Wrekin England **37** E2 52 47N 2 22W
Newport Wales **36** D1 51 35N 3 00W
Newquay England **40** B3 50 25N 5 05W
New Quay Wales **36** C2 52 13N 4 22W
Newry Northern Ireland **32** C2 54 11N 6 20W
Newry and Mourne district Northern Ireland **32** C2 54 10N 6 20W
New Siberian Islands Russia **57** Q5 75 00N 145 00E
New South Wales state Australia **72** E2 32 00S 145 00E
Newton Abbot England **40** D3 50 32N 3 36W

Newton Aycliffe England **34** D3 54 37N 1 34W
Newtonmore Scotland **31** E2 57 04N 4 08W
Newton Stewart Scotland **33** E2 54 57N 4 29W
Newtown Wales **36** D2 52 32N 3 19W
Newtownabbey Northern Ireland **32** D2 54 40N 5 54W
Newtownabbey district Northern Ireland **32** D2 54 45N 5 55W
Newtownards Northern Ireland **32** D2 54 36N 5 41W
Newtownhamilton Northern Ireland **32** C2 54 12N 6 35W
Newtownstewart Northern Ireland **32** B2 54 43N 7 24W
New York USA **79** M5 40 40N 73 50W
New York state USA **79** M5 43 00N 76 00W
NEW ZEALAND 73
Ngaoundére Cameroon **66** D3 7 20N 13 35E
Nguru Nigeria **66** D4 12 53N 10 30E
Nha-Trang Vietnam **60** B4 12 15N 109 10E
Niamey Niger **66** C4 13 32N 2 05E
NICARAGUA 79 L2
Nicaragua, Lake Nicaragua **79** L2 11 50N 86 00W
Nice France **48** G1 43 42N 7 16E
Nicobar Islands India **58** E1 8 30N 94 00E
Nicosia Cyprus **43** N3 35 11N 33 23E
Nidd r. England **34** D3 54 05N 1 40W
Nieuwpoort Belgium **39** G2 51 08N 2 45E
NIGER 66 C4
Niger r. Africa **66** B4 11 00N 5 00W
NIGERIA 66 C3
Niigata Japan **59** M2 37 58N 139 02E
Nijmegen Netherlands **49** C4 51 50N 5 52E
Nile r. Sudan/Egypt **61** D3 28 00N 31 00E
Nimba, Mount Guinea/Liberia **66** B3 7 39N 8 30W
Nîmes France **48** F1 43 50N 4 21E
Niort France **48** C3 46 19N 0 27W
Niš Serbia and Montenegro **42** L4 43 20N 21 54E
Niteroi Brazil **84** E5 22 54S 43 06W
Nith r. Scotland **33** F3 55 15N 3 50W
Nizhniy Novgorod Russia **56** E3 56 20N 44 00E
Nizhniy Tagil Russia **56** F3 58 00N 59 58E
Norfolk USA **79** M4 36 54N 76 18W
Norfolk co. England **39** E3/F3 52 45N 1 00E
Norfolk Broads geog. reg. England **39** F3 52 45N 1 30E
Norilsk Russia **57** J4 69 21N 88 02E
Normandy geog. reg. France **48** C4/D4 49 00N 0 00
Norrköping Sweden **42** K7 58 35N 16 10E
Northallerton England **35** D3 54 20N 1 26W
Northampton England **38** D3 52 14N 0 54W
Northamptonshire co. England **38** C3/D3 52 20N 1 00W
North Ayrshire u.a. Scotland **32/33** D3/E3 55 30N 5 05W
North Berwick Scotland **33** G4 56 04N 2 44W
North Cape New Zealand **73** C4 34 25S 173 03E
North Carolina state USA **79** M4 36 00N 80 00W
North Channel British Isles **32** D2/3 55 20N 5 50W
North Dakota state USA **79** J5 47 00N 102 00W
North Down district Northern Ireland **32** D2 54 35N 5 50W
North Downs hills England **38/39** D2/E2 51 15N 0 30E
North Dvina r. Russia **56** E4 63 00N 43 00E
North East Lincolnshire u.a. England **35** E2 53 30N 0 10W
Northern Ireland province United Kingdom **32** B2 54 40N 7 00W
NORTHERN MARIANAS 5
Northern Territory Australia **72** D4 19 00S 132 00E
North Esk r. Scotland **31** G1 56 50N 2 50W
North Foreland c. England **39** F2 51 23N 1 27E
North Friesian Islands Germany **49** E6 54 00N 8 00E
North Island New Zealand **73** D3 39 00N 176 00E
NORTH KOREA 59 D4
North Lanarkshire u.a. Scotland **33** F3 55 50N 3 55W
North Lincolnshire u.a. England **35** E2 53 40N 0 40W
North Pole Arctic Ocean **86** 90 00N
North Ronaldsay i. Scotland **29** C2 59 23N 2 26W
North Sea Atlantic Ocean **42** G7 57 00N 4 00E
North Shields England **35** D4 55 01N 1 26W
North Uist i. Scotland **30** B2 57 04N 7 15W
Northumberland co. England **33** G3/H3 55 10N 2 00W
Northwest Highlands Scotland **30/31** D2/E3 58 00N 5 00W
North West Somerset u.a. England **38** B2 51 25N 2 50W
Northwest Territories territory Canada **78** H7 64 00N 110 30W
Northwich England **34** C2 53 16N 2 32W
North York Moors England **35** E3 54 22N 0 45W
North Yorkshire co. England **34/35** D3/E3 54 05N 1 20W
Norton England **35** E3 54 08N 0 48W
NORWAY 42 H7/M9
Norwich England **39** F3 52 38N 1 18E
Nottingham England **37** F2 52 58N 1 10W
Nottingham City u.a. England **37** F2 52 58N 1 10W
Nottinghamshire co. England **37** F3/G3 53 20N 1 00W
Nouadhibou Mauritania **66** A5 20 54N 17 01W
Nouakchott Mauritania **66** A4 18 09N 15 58W
Novara Italy **51** B6 45 27N 8 37E
Nova Scotia province Canada **79** N5 45 30N 63 00W
Novaya Zemlya is. Russia **56** F5 74 00N 55 00E
Novgorod Russia **42** N7 58 30N 31 20E
Novokuznetsk Russia **57** J3 53 45N 87 12E
Novosibirsk Russia **56** J3 55 04N 83 05E
Nuevo Laredo Mexico **79** K3 27 39N 99 30W
Nu Jiang (Salween) r. China/Myanmar **58** E4 31 30N 94 00E
Nullarbor Plain Australia **72** C2 30 30S 129 00E
Nunavut territory Canada **78** K7 67 30N 95 30W
Nuneaton England **37** F2 52 32N 1 28W
Nunivak i. USA **78** C6 60 00N 166 00W
Nuremberg Germany **49** F3 49 27N 11 05E
Nuuk (Godthåb) Greenland **78** P7 64 10N 51 40W
Nyala Sudan **61** C2 12 01N 24 50E
Nyasa, Lake (Lake Malawi) Malawi/Mozambique **67** D3 12 00S 35 00E

O

Oahe, Lake USA **79** J5 45 00N 100 00W
Oamaru New Zealand **73** C1 45 06S 170 58E
Ob r. Russia **56** G4 64 00N 65 00E
Oban Scotland **30** D1 56 25N 5 29W
Oberammergau Germany **49** F2 47 35N 11 07E
Obihiro Japan **59** N3 42 56N 143 10E
Obi Islands i. Indonesia **60** D2 1 30S 127 30E
Ochil Hills Scotland **31** F1 56 15N 3 30W
Odense Denmark **42** J7 55 24N 10 25E
Oder r. Europe **49** H5 53 00N 14 30E
Odessa Ukraine **56** D2 46 30N 30 46E
Ogbomoso Nigeria **66** C3 8 05N 4 11E
Ohio r. USA **79** L4 37 00N 88 00W
Ohio state USA **79** L5 40 00N 83 00W
Oiapoque Brazil **84** D8 3 54N 51 46W
Oise r. France **48** E4 49 30N 3 00E
Oita Japan **59** L1 33 15N 131 36E
Okavango Swamp Botswana **67** C3 19 00S 23 00E
Okayama Japan **59** L1 34 40N 133 54E
Okehampton England **40** C3 50 44N 4 00W
Okhotsk, Sea of Russia **57** Q3 56 00N 145 00E
Oki Islands Japan **59** L2 36 05N 133 00E
Oklahoma state USA **79** K4 35 00N 97 00W
Oklahoma City USA **79** K4 35 28N 97 33W
Öland i. Sweden **42** K7 56 45N 51 50E
Olbia Italy **51** B4 40 56N 9 30E
Oldenburg Germany **49** E5 53 08N 8 13E
Oldham England **34** C2 53 33N 2 07W
Oldmeldrum Scotland **31** G2 57 20N 2 20W
Olongapo The Philippines **60** D4 14 49N 120 17E
Olympus, Mt. Greece **42** L4 40 05N 22 21E
Omagh Northern Ireland **32** B2 54 36N 7 18W
Omagh district Northern Ireland **32** B2 54 30N 7 30W
Omaha USA **79** K5 41 15N 96 00W
OMAN 61 F3
Oman, Gulf of Iran/Oman **61** F3 24 30N 58 30E
Omdurman Sudan **61** D2 15 37N 32 29E
Omsk Russia **56** H3 55 00N 73 22E
Onega, Lake l. Russia **56** D4 62 00N 40 00E
Onitsha Nigeria **66** C3 6 10N 6 47E
Ontario province Canada **79** L6 51 00N 85 00W
Ontario, Lake Canada/USA **79** M5 43 45N 78 00W
Oostende Belgium **49** B4 51 13N 2 55E
Oporto Portugal **50** A4 41 09N 8 37W
Oran Algeria **66** B6 35 45N 0 38W
Orange France **48** F2 44 08N 4 48E
Orange r. Southern Africa **67** B2 29 00S 19 00E
Ord r. Australia **72** C4 15 00S 129 00E
Örebro Sweden **42** K7 59 17N 15 13E
Oregon state USA **79** G5 44 00N 120 00W
Orel Russia **56** D3 52 58N 36 04E
Orenburg Russia **56** F3 51 50N 55 00E
Orford Ness c. England **39** F3 52 05N 1 34E
Orkney Islands u.a. Scotland **29** B1/C2 59 00N 3 00W
Orléans France **48** D3 47 54N 1 54E
Orne r. France **41** G1/2 49 00N 0 00
Oronsay i. Scotland **32** C4 56 00N 6 15W
Orsk Russia **56** F3 51 13N 58 35E
Ortles mt. Italy **49** F4 46 31N 10 33E
Oruro Bolivia **84** C6 17 59S 67 08W
Osaka Japan **59** M1 34 40N 135 30E
Osh Kyrgyzstan **56** H2 40 37N 72 49E
Oslo Norway **42** J7 59 56N 10 45E
Osnabrück Germany **49** E5 52 17N 8 03E
Osorno Chile **85** B3 40 35S 73 14W
Ostrava Czech Republic **42** K5 49 50N 18 15E
Oswestry England **36** D2 52 52N 3 03W
Otranto Italy **51** G4 40 08N 18 30E
Ottawa Canada **79** M5 46 00N 77 00W
Ouagadougou Burkina **66** B4 12 20N 1 40W
Oubangui r. Central Africa **67** C5 4 00N 21 00E
Oujda Morocco **66** B6 34 41N 1 45W
Oulu Finland **42** M9 65 02N 25 27E
Oundle England **38** D3 52 29N 0 29W
Ouse r. England **35** D3 54 00N 1 05W
Outer Hebrides is. Scotland **30** B1/C3 58 00N 7 00W
Out Skerries is. Scotland **29** E3 60 25N 0 46W
Oviedo Spain **50** C5 43 21N 7 18W
Oxford England **38** C2 51 46N 1 15W
Oxfordshire co. England **38** C2 51 50N 1 25W

P

Pabbay i. Scotland **30** B2 57 47N 7 14W
Pacific Ocean 79 J2
Padang Indonesia **60** B2 1 00S 100 21E
Paderborn Germany **49** E4 51 43N 8 44E
Padstow England **40** C3 50 33N 4 56W
Padua Italy **51** C6 45 24N 11 53E
Paisley Scotland **33** E3 55 50N 4 26W
PAKISTAN 58 B3
Palau Islands Pacific Ocean **60** E3 7 30N 134 30E
Palawan i. The Philippines **60** C3 10 00N 119 00E
Palembang Indonesia **60** B2 2 59S 104 45E
Palermo Italy **51** D3 38 08N 13 23E
Palma Balearic Islands **50** G3 39 35N 2 39E
Palmerston North New Zealand **73** D2 40 22S 175 37E
Pamirs mts. Asia **58** C4 38 00N 74 00E
Pamplona Spain **50** E5 42 49N 1 39W
PANAMA 79 L1
Panamá Panama **79** M1 8 57N 19 30W
Panama Canal Panama **79** M1 9 00N 80 00W
Panay i. The Philippines **60** D4 11 00N 122 00E
Pantelleria i. Italy **51** D2 36 00N 12 00E
Papa Stour i. Scotland **29** D3 60 20N 1 42W
Papa Westray i. Scotland **29** C2 59 22N 2 54W
PAPUA NEW GUINEA 72 E5
PARAGUAY 84 D5
Paraguay r. South America **84** D5 21 20S 58 20W
Paramaribo Suriname **84** D8 5 52N 55 14W

Paraná Argentina 85 C4 31 45S 60 30W
Paraná r. South America 85 D5 30 00S 59 50W
Paris France 48 E4 48 52N 2 20E
Parma Italy 51 C6 44 48N 10 19E
Parnaíba Brazil 84 E7 2 58S 41 46W
Parrett r. England 41 E4 51 05N 2 55W
Parry Islands Canada 78 J8 75 15N 109 00W
Pasto Colombia 84 B8 1 12N 77 17W
Patagonia geog. reg. Argentina 85 C3 48 00S 70 00W
Patna India 58 D3 25 37N 85 12E
Patra Greece 42 L3 38 14N 21 44E
Pau France 48 C1 43 18N 0 22W
Peace r. Canada 78 H6 57 30N 117 00W
Pechora Russia 43 T9 65 14N 57 18E
Pechora r. Russia 43 S9 66 00N 53 00E
Peebles Scotland 33 F3 55 39N 3 12W
Peel Isle of Man British Isles 33 E2 54 14N 4 42W
Peel Fell sum. Scotland/England 33 G3 55 17N 2 35W
Pegasus Bay New Zealand 73 C2 43 22S 172 55E
Pegu Myanmar 58 E2 17 18N 96 31E
Peipus, Lake Estonia/Russia 42 M7 58 00N 27 45E
Peloponnese geog. reg. Greece 42 L3 37 00N 22 00E
Pelotas Brazil 85 D4 31 45S 52 20W
Pematangsiantar Indonesia 60 A3 2 59N 99 01E
Pembroke Wales 36 C1 51 41N 4 55W
Pembrokeshire u.a. Wales 36 B1/C1 51 50N 5 00W
Peñalara mt. Spain 50 D4 40 51N 3 58W
Penicuik Scotland 33 F3 55 50N 3 14W
Pennines hills England 34 C3 54 30N 2 10W
Pennsylvania state USA 79 M5 41 00N 78 00W
Penrith England 34 C2 53 46N 2 44W
Penryn England 40 B3 50 09N 5 06W
Pentland Firth sd. Scotland 29 B1 58 45N 3 10W
Pentland Hills Scotland 33 F3 55 45N 3 30W
Pen-y-ghent sum. England 34 C3 54 10N 2 14W
Penza Russia 56 F3 53 11N 45 00E
Penzance England 40 B3 50 07N 5 33W
Perm Russia 56 F3 58 01N 56 10E
Perpignan France 48 E1 42 42N 2 54E
Perth Australia 72 B2 31 57S 115 52E
Perth Scotland 31 F2 56 42N 3 28W
Perth and Kinross u.a. Scotland 31 E1/F1 56 45N 3 50W
PERU 84 B6
Perugia Italy 51 D5 43 07N 12 23E
Pescara Italy 51 E5 42 27N 14 13E
Peshawar Pakistan 58 C4 34 01N 71 33E
Peterborough England 38 D3 52 35N 0 15W
Peterborough u.a. England 38 D3 52 40N 0 15W
Peterhead Scotland 31 H2 57 30N 1 46W
Peterlee England 35 D3 54 46N 1 19W
Petersfield England 38 D2 51 00N 0 56W
Petropavlovsk Kazakhstan 56 H3 54 53N 69 13E
Petropavlovsk-Kamchatskiy Russia 57 R3 53 03N 158 43E
Petrozavodsk Russia 42 N8 61 46N 34 19E
Pforzheim Germany 49 E3 48 53N 8 41E
Philadelphia USA 79 M4 40 00N 75 10W
PHILIPPINES 60 D3/4
Phnom Penh Cambodia 60 B4 11 35N 104 55E
Phoenix USA 79 H4 33 30N 112 03W
Piacenza Italy 51 B6 45 03N 9 41E
Pickering England 35 E3 54 14N 0 46W
Pickering, Vale of England 35 E3 54 10N 0 45W
Pietermaritzburg Republic of South Africa 67 D2 29 36S 30 24E
Pilcomayo r. South America 84 C5 24 00S 61 00W
Piraeus Greece 42 L3 37 57N 23 42E
Pisa Italy 51 C5 43 43N 10 24E
Pitcairn Islands Pacific Ocean 4 25 04S 130 06W
Pitlochry Scotland 31 F1 56 43N 3 45W
Piura Peru 84 A7 5 15S 80 38W
Plenty, Bay of New Zealand 73 D3 37 48S 177 12E
Ploiești Romania 42 M4 44 57N 26 01E
Plovdiv Bulgaria 42 L4 42 08N 24 45E
Plymouth England 40 C3 50 23N 4 10W
Plymouth u.a. England 40 C3 50 24N 4 10W
Plynlimon mt. Wales 36 D2 52 28N 3 47W
Po r. Italy 51 C6/D6 45 00N 12 00E
Pobeda Peak Kyrgyzstan 56 J2 42 25N 80 15E
Pointe Noire tn. Congo 66 D2 4 46S 11 53E
Poitiers France 48 D3 46 35N 0 20E
POLAND 42 K6
Ponta Grossa Brazil 85 D5 25 07S 50 09W
Pontardulais Wales 36 C1 51 43N 4 02W
Pontefract England 35 D2 53 42N 1 18W
Pontianak Indonesia 60 B2 0 05S 109 16E
Pontypool Wales 36 D1 51 43N 3 02W
Pontypridd Wales 36 D1 51 37N 3 22W
Poole England 41 F3 50 43N 1 59W
Poole u.a. England 41 F3 50 45N 1 58W
Poolewe Scotland 30 D2 57 45N 5 37W
Poopo, Lake Bolivia 84 C6 18 30S 67 20W
Popocatepetl vol. Mexico 79 K2 19 02N 98 38W
Porbandar India 58 B3 21 04N 69 40E
Portadown Northern Ireland 32 C2 54 26N 6 27W
Port Askaig Scotland 32 C3 55 51N 6 07W
Port Augusta Australia 72 D2 32 30S 137 48E
Port-au-Prince Haiti 79 M2 18 33N 72 20W
Port Elizabeth Republic of South Africa 67 C1 33 58S 25 36E
Port Ellen Scotland 32 C3 55 39N 6 12W
Port Gentil Gabon 66 C2 0 40S 8 50E
Port Glasgow Scotland 33 E3 55 56N 4 41W
Port Harcourt Nigeria 66 C3 4 43N 7 05E
Port Hedland Australia 72 B3 20 18S 118 35E
Porthmadog Wales 36 C2 52 55N 4 08W
Portknockie Scotland 31 G2 57 42N 2 52W
Portland USA 78 G5 45 32N 122 40W
Portland Bill c. England 41 E3 50 31N 2 27W
Port Lincoln Australia 72 D3 34 44S 135 52E
Port Moresby Papua New Guinea 72 E5 9 30S 147 07E
Portnahaven Scotland 32 C3 55 41N 6 31W
Port of Ness Scotland 30 C3 58 29N 6 13W
Port of Spain Trinidad and Tobago 79 N2 10 38N 61 31W

Porto Novo Benin 66 C3 6 30N 2 47E
Pörto Velho Brazil 84 C7 8 45S 63 54W
Portpatrick Scotland 32 D2 54 51N 5 07W
Port Pirie Australia 72 D2 33 10S 138 02E
Portree Scotland 30 C2 57 24N 6 12W
Portrush Northern Ireland 32 C3 55 12N 6 40W
Port Said Egypt 61 D4 31 17N 32 18E
Portsmouth England 38 C1 50 48N 1 05W
Portsmouth u.a. England 38 C1 50 50N 1 04W
Portsoy Scotland 31 G2 57 41N 2 41W
Portstewart Northern Ireland 32 C3 55 11N 6 43W
Port Sudan Sudan 61 D2 19 38N 37 07E
Port Talbot Wales 36 D1 51 36N 3 47W
PORTUGAL 50 A3
Potenza Italy 51 E4 40 38N 15 48E
Potsdam Germany 49 G5 52 24N 13 04E
Powys u.a. Wales 36 D2 52 10N 3 00E
Poyang Hu l. China 59 C2 29 00N 116 30E
Poznań Poland 42 K6 52 25N 16 53E
Prague Czech Republic 42 J6 50 06N 14 26E
Prato Italy 51 C5 43 53N 11 06E
Preston England 34 C2 53 46N 2 42W
Prestwick Scotland 33 E3 55 30N 4 37W
Pretoria Republic of South Africa 67 C2 25 45S 28 12E
Prijedor Bosnia-Herzegovina 51 F6 45 00N 16 41E
Prince Edward Island province Canada 79 N5 46 30N 63 00W
Prince Rupert Canada 78 F6 54 09N 130 20W
Principe i. Gulf of Guinea 66 C3 1 37N 7 27E
Pripet r. Belarus 42 M6 52 00N 28 00E
Provence geog. reg. France 48 F1/G1 44 00N 6 00E
Providence USA 79 M5 41 50N 71 28W
Providence, Cape New Zealand 73 B1 46 01S 166 28E
Prudhoe Bay tn. USA 78 E7 70 05N 148 20W
Pskov Russia 42 M7 57 48N 28 26E
Pucallpa Peru 84 B7 8 21S 74 33W
Puebla Mexico 79 K2 19 03N 98 10W
Puerto Montt Chile 85 B3 41 28S 73 00W
PUERTO RICO 79 N2
Puig Mayor mt. Balearic Islands 50 G3 39 49N 2 47E
Pukekohe New Zealand 73 D3 37 12S 174 54E
Pune India 58 C2 18 34N 73 58E
Punta Arenas Chile 85 B2 53 10S 70 56W
Pusan South Korea 59 D3 35 05N 129 02E
Puy de Sancy mt. France 48 E2 45 32N 2 48E
Pwllheli Wales 36 C2 52 53N 4 25W
Pyongyang North Korea 59 D3 39 00N 125 47E
Pyrénées mts. France/Spain 50 F5 42 50N 0 30E

Q

Qaidam Pendi (Qaidam Basin) China 59 A4 37 30N 94 00E
QATAR 61 F3
Qattara Depression Egypt 61 C3 24 00N 27 30E
Qazvin Iran 61 E4 36 16N 50 00E
Qingdao China 59 D3 36 04N 120 22E
Qinghai Hu l. China 59 B3 37 00N 100 00E
Qiqihar China 59 D4 47 23N 124 00E
Qizil r. Iran 61 E4 36 00N 48 00E
Qom Iran 61 F4 34 39N 50 57E
Quantock Hills England 41 D4 51 05N 3 15W
Québec Canada 79 M5 46 50N 71 15W
Québec province Canada 79 M6 51 00N 75 00W
Queen Charlotte Islands Canada 78 F6 53 30N 131 50W
Queen Elizabeth Islands Canada 78 J8 77 30N 105 00W
Queensland state Australia 72 E3 22 00S 143 00E
Queenstown New Zealand 73 B1 45 02S 168 40E
Quetta Pakistan 58 B4 30 12N 67 00E
Quezon City The Philippines 60 D4 14 39N 121 02E
Qui Nhon Vietnam 60 B4 13 47N 109 11E
Quito Ecuador 84 B7 0 14S 78 30W

R

Raasay i. Scotland 30 C2 57 25N 6 05W
Rabat-Salé Morocco 66 B6 34 02N 6 51W
Rainier, Mount USA 78 G5 46 25N 121 45W
Raipur India 58 D3 21 16N 81 42E
Rajang r. Malaysia 60 C3 2 00N 114 00E
Rakaia r. New Zealand 73 C2 43 00S 172 00E
Ramsey Isle of Man British Isles 33 E2 54 19N 4 23W
Ramsgate England 39 F2 51 20N 1 25E
Randalstown Northern Ireland 32 C2 54 45N 6 19W
Rangitaiki r. New Zealand 73 D3 38 45S 177 00E
Ras Dashen mt. Ethiopia 61 D2 13 15N 38 27E
Rasht Iran 61 E4 37 18N 49 38E
Rathlin Island Northern Ireland 32 C3 55 20N 6 10W
Rathlin Sound Northern Ireland 32 C3 55 20N 6 25W
Rattray Scotland 31 F1 56 36N 3 20W
Ravenna Italy 51 D6 44 25N 12 12E
Rawalpindi Pakistan 58 C4 33 36N 73 04E
Reading England 38 D2 51 28N 0 59W
Reading u.a. England 38 D2 51 27N 0 59W
Recife Brazil 84 F7 8 06S 34 53W
Recklinghausen Germany 49 D4 51 37N 7 11E
Redbridge England 39 E2 51 34N 0 05E
Redcar England 35 D3 54 37N 1 04W
Redcar and Cleveland u.a. England 35 D3/E3 54 30N 1 00W
Redditch England 37 F2 52 19N 1 56W
Rede r. England 33 G3 55 15N 2 20W
Redhill tn. England 38 D2 51 14N 0 11W
Redruth England 40 B3 50 13N 5 14W
Red Sea Middle East 61 D2/3 20 00N 38 00E
Regensburg Germany 49 G3 49 01N 12 07E
Reggio di Calabria Italy 51 F3 38 06N 15 39E
Reggio nell'Emilia Italy 51 C6 44 42N 10 37E
Regina Canada 78 J6 50 30N 104 38W
Reigate England 38 D2 51 14N 0 13W
Reims France 48 F4 49 15N 4 02E
Remscheid Germany 49 D4 51 10N 7 11E
Renfrewshire u.a. Scotland 33 E3 55 50N 4 40W
Rennes France 48 C4 48 06N 1 40W
Reno USA 79 H4 39 32N 119 49W

REPUBLIC OF IRELAND 28
REPUBLIC OF SOUTH AFRICA 67 C2
Resistencia Argentina 85 D5 27 28S 59 00W
Reutlingen Germany 49 E3 48 30N 9 13E
Reykjavik Iceland 42 B8 64 09N 21 58W
Rhayader Wales 36 D2 52 18N 3 30W
Rhine r. Europe 49 D4 50 30N 7 30E
Rhode Island state USA 79 M5 41 00N 71 00W
Rhodes i. Greece 42 M3 36 00N 28 00E
Rhondda Wales 36 D1 51 40N 3 30W
Rhondda Cynon Taff u.a. Wales 36 D1 51 40N 3 30W
Rhône r. Switzerland/France 48 F2 45 00N 4 50E
Rhum i. Scotland 30 C1/2 57 00N 6 20W
Rhyl Wales 36 D3 53 19N 3 29W
Ribble r. England 34 C2 53 45N 2 30W
Ribeirão Preto Brazil 84 E5 21 09S 47 48W
Richmond England 34 D3 54 24N 1 44W
Richmond USA 79 M4 37 34N 77 27W
Richmond upon Thames England 38 D2 51 28N 0 19W
Riga Latvia 42 L7 56 53N 24 08E
Rimini Italy 51 D6 44 03N 12 34E
Ringwood England 41 F3 50 51N 1 47W
Rio Branco tn. Brazil 84 C7 9 59S 67 49W
Rio de Janeiro tn. Brazil 84 E5 22 53S 43 17W
Rio Grande r. Mexico/USA 79 J3 29 30N 105 00W
Rio Grande tn. Brazil 85 D4 32 03S 52 08W
Ripon England 34 D3 54 08N 1 31W
River Plate Estuary Uruguay/Argentina 85 D4 35 00S 57 00W
Riyadh Saudi Arabia 61 E3 24 39N 46 46E
Roanne France 48 F3 46 02N 4 05E
Robson, Mount Canada 78 H6 53 08N 118 18W
Rochdale England 34 C2 53 38N 2 09W
Rochester England 39 E2 51 24N 0 30E
Rochester USA 79 M5 43 12N 77 37W
Rockhampton Australia 72 F3 23 22S 150 30E
Rocky Mountains Canada/USA 78/79 F7/J4 46 00N 110 00W
Roebourne Australia 72 B3 20 45S 117 08E
ROMANIA 42 L5
Rome Italy 51 D4 41 53N 12 30E
Romsey England 38 C1 50 59N 1 30W
Ronne Ice Shelf Antarctica 86 77 00S 60 00W
Rora Head Scotland 29 B1 58 52N 3 26W
Roraima mt. Guyana 84 C8 5 14N 60 44W
Rosario Argentina 84 C4 33 00S 60 40W
Rosehearty Scotland 31 G2 57 42N 2 07W
Rosenheim Germany 51 E5 47 51N 12 09E
Rossano Italy 51 F3 39 35N 16 38E
Ross Ice Shelf Antarctica 86 80 00S 180 00
Ross of Mull pen. Scotland 30 C1 56 20N 6 10W
Ross-on-Wye England 37 E1 51 55N 2 35W
Ross Sea Antarctica 86 75 00S 180 00
Rostock Germany 49 G6 54 06N 12 09E
Rostov-on-Don Russia 56 E2 47 15N 39 45E
Rotherham England 35 D2 53 26N 1 20W
Rothes Scotland 31 F2 57 31N 3 13W
Rothesay Scotland 32 D3 55 51N 5 03W
Rotorua New Zealand 73 D3 38 08S 176 14E
Rotterdam Netherlands 49 C4 51 54N 4 28E
Rouen France 48 D4 49 26N 1 05E
Rousay i. Scotland 29 B2 59 10N 3 00W
Royal Leamington Spa England 37 F2 52 18N 1 31W
Royal Tunbridge Wells England 39 E2 51 08N 0 16E
Royston England 38 D3 52 03N 0 01W
Ruapehu mt. New Zealand 73 D3 39 18S 175 34E
Rubha Hunish c. Scotland 30 C2 57 40N 6 20W
Rugby England 37 F2 52 23N 1 15W
Rugeley England 37 F2 52 46N 1 55W
Rügen i. Germany 49 G6 54 00N 14 00E
Rukwa, Lake Tanzania 67 C4 8 00S 32 00E
Runcorn England 34 C2 53 20N 2 44W
RUSSIA see RUSSIAN FEDERATION
RUSSIAN FEDERATION (RUSSIA) 56/57
Rutland u.a. England 38 D3 52 42N 0 35W
Rutland Water l. England 38 D3 52 40N 0 37W
Ruvuma r. Tanzania/Mozambique 67 D3 11 00S 39 00E
Ruwenzori, Mount Congo Dem. Rep./Uganda 67 C5 0 23N 29 54E
RWANDA 67 C4
Ryazan Russia 56 D3 54 37N 39 43E
Rybinsk Russia 56 D3 58 03N 38 50E
Rybinsk Reservoir Russia 43 P7 59 00N 38 00E
Ryde England 38 C1 50 44N 1 10W
Rye England 39 E1 50 57N 0 44E
Ryukyu Islands Japan 59 D2 27 30N 127 30E

S

Saale r. Germany 49 F4 51 15N 12 00E
Saar r. Germany/France 49 D3 49 15N 6 30E
Saarbrücken Germany 49 D3 49 15N 6 58E
Sabadell Spain 50 G4 41 33N 2 07E
Sabah admin. Malaysia 60 C3 5 00N 115 00E
Sabhā Libya 66 D5 27 02N 14 26E
Sacramento USA 79 G4 38 32N 121 30W
Sado i. Japan 59 M2 38 20N 138 30E
Saffron Walden England 39 E3 52 01N 0 15E
Sahara Desert North Africa 66 A5/E5 22 50N 0 00
St. Abb's Head c. Scotland 33 G3 55 55N 2 09W
St. Agnes England 40 B3 50 18N 5 13W
St. Albans England 38 D2 51 46N 0 21W
St. Albans Head c. England 41 E3 50 34N 2 04W
St. Andrews Scotland 33 G4 56 20N 2 48W
St. Anton Austria 49 F2 47 09N 10 17E
St. Austell England 40 C3 50 20N 4 48W
St. Bees Head England 34 B3 54 31N 3 39W
St. Bride's Bay Wales 36 B1 51 50N 5 15W
St. Brieuc France 48 B4 48 31N 2 45W
St. Catherine's Point England 38 C1 50 34N 1 18W
St. David's Wales 36 B1 51 54N 5 16W
St. David's Head Wales 36 B1 51 55N 5 19W
St-Denis France 48 E4 48 57N 2 22E
St-Dizier France 48 F4 48 38N 4 58E

Saintes France 48 C2 45 44N 0 38W
St-Étienne France 48 F2 45 26N 4 23E
St. Gallen Switzerland 51 B7 47 25N 9 23E
St. George's Channel British Isles 36 A1/B2 52 00N 6 00W
St. Germain France 48 E4 48 55N 2 08E
St. Gotthard Pass Switzerland 49 E2 46 34N 8 31E
St. Helena i. Atlantic Ocean 5 15 58S 5 43W
St. Helens England 34 C2 53 28N 2 44W
St. Helier Jersey Channel Islands 41 E2 49 12N 2 07W
St. Ives Cambridgeshire England 38 D3 52 20N 0 05W
St. Ives Cornwall England 40 B3 50 12N 5 29W
Saint John Canada 79 N5 45 16N 66 03W
St. John's Canada 79 P5 47 34N 52 43W
St. John's Point Northern Ireland 32 D2 54 15N 5 40W
St. John's Town of Dalry Scotland 33 E3 55 07N 4 10W
St. Just England 40 B3 50 07N 5 41W
St. Kilda i. Scotland 42 A7 57 49N 8 34W
ST. KITTS AND NEVIS 79 N2
St. Lawrence i. Alaska 78 C7 63 15N 169 50W
St. Lawrence r. Canada/USA 79 M5 47 30N 70 00W
St. Lawrence, Gulf of Canada 79 N5 49 00N 62 30W
St-Lô France 41 F2 49 07N 1 05W
St. Louis Senegal 66 A4 16 01N 16 30W
St. Louis USA 79 K4 38 40N 90 15W
ST. LUCIA 79 N2
St. Magnus Bay Scotland 29 D3 60 25N 1 35W
St-Malo France 48 C4 48 39N 2 00W
St. Martin's i. England 40 A2 49 58N 6 17W
St. Mary's i. England 40 A2 49 55N 6 18W
St. Mary's Loch l. Scotland 33 F3 55 29N 3 12W
St. Matthew i. USA 78 B7 60 30N 172 30W
St. Moritz Switzerland 49 E2 46 30N 9 51E
St-Nazaire France 48 B3 47 17N 2 12W
St. Neots England 38 D3 52 14N 0 17W
St-Omer France 39 G1 50 45N 2 15E
St. Paul USA 79 K5 45 00N 93 10W
St. Peter-Port Guernsey Channel Islands 41 E2 49 27N 3 32W
St. Petersburg Russia 56 D3 59 55N 30 25E
Saint Pierre & Miquelon is. Atlantic Ocean 79 P5 47 00N 56 20W
St-Quentin France 48 E4 49 51N 3 17E
St-Tropez France 48 G1 43 16N 6 39E
ST. VINCENT AND THE GRENADINES 79 N2
Sakhalin i. Russia 57 Q3 50 00N 143 00E
Salado r. Argentina 85 C5 28 30S 62 30W
Salalah Oman 61 F2 17 00N 54 04E
Salamanca Spain 50 C4 40 58N 5 40W
Salcombe England 40 D3 50 13N 3 47W
Sale England 34 C2 53 26N 2 19W
Salem India 58 C2 11 38N 78 08E
Salerno Italy 51 E4 40 40N 14 46E
Salford England 34 C2 53 30N 2 16W
Salisbury England 38 C2 51 05N 1 48W
Salisbury Plain England 38 C2 51 10N 1 55W
Salta Argentina 84 C5 24 46S 65 28W
Saltash England 40 C3 50 24N 4 12W
Saltcoats Scotland 33 E3 55 38N 4 47W
Saltillo Mexico 79 J3 25 30N 101 00W
Salt Lake City USA 79 H5 40 45N 111 55W
Salvador Brazil 84 F6 12 58S 38 29W
Salween r. see Nu Jiang
Salzburg Austria 42 J5 47 48N 13 03E
Salzgitter Germany 49 F5 52 13N 10 20E
Samar i. The Philippines 60 D4 12 00N 125 00E
Samara Russia 56 F3 53 10N 50 10E
Samarinda Indonesia 60 C2 0 30S 117 09E
Samarkand Uzbekistan 56 G1 39 40N 66 57E
Sambre r. France 48 E5 50 15N 3 50E
SAMOA 4/5
Samsun Turkey 43 P4 41 17N 36 22E
Sana Yemen Republic 61 E2 15 23N 44 14E
Sanaga r. Cameroon 66 D3 5 00N 14 00E
San Antonio USA 79 K3 29 00N 97 00W
San Antonio Abad Balearic Islands 50 F3 38 59N 1 19E
San Benito Islands Mexico 79 H3 28 00N 115 00W
San Cristobal Venezuela 84 B8 7 46N 72 15W
Sandakan Malaysia 60 C3 5 52N 118 04E
Sanday i. Scotland 29 C2 59 15N 2 30W
San Diego USA 79 H4 32 45N 117 10W
Sandown England 38 C1 50 39N 1 09W
San Francisco USA 79 G4 37 45N 122 27W
San Francisco Javier Balearic Islands 50 F3 38 43N 1 26E
San José Costa Rica 79 L1 9 59N 84 04W
San Juan Argentina 85 C4 31 33S 68 31W
San Juan Puerto Rico 79 N2 18 29N 66 08W
Sanliurfa Turkey 43 P3 37 09N 38 48E
San Luis Potosí Mexico 79 J3 22 10N 101 00W
SAN MARINO 51 D5
San Miguel de Tucumán Argentina 85 C5 26 47S 65 15W
San Pedro Sula Honduras 79 L2 15 26N 88 01W
Sanquhar Scotland 33 F3 55 22N 3 56W
San Salvador El Salvador 79 L2 13 40N 89 10W
San Sebastián Spain 50 E5 43 19N 1 59W
Santa Clara Cuba 79 M3 22 25N 79 58W
Santa Cruz Bolivia 84 C6 17 50S 63 10W
Santa Fé Argentina 84 C4 31 35S 60 50W
Santa Marta Colombia 84 B9 11 18N 74 10W
Santander Spain 50 D5 43 28N 3 48W
Santañy Balearic Islands 50 G3 39 22N 3 07E
Santarém Brazil 84 D7 2 26S 54 41W
Santarém Portugal 50 A3 39 14N 8 40W
Santiago Chile 85 B4 33 30S 70 40W
Santiago r. Mexico 79 J3 104 00W 21 30N
Santiago de Cuba Cuba 79 M3 20 00N 75 49W
Santiago del Estero Argentina 85 C5 27 47S 64 15W
Santo Domingo Dominican Republic 79 N2 18 30N 69 57W
Santos Brazil 84 E5 23 56S 46 22W
São Francisco r. Brazil 84 E6 13 40S 44 00W

São Luís Brazil **84** E7 2 34S 44 16W
Saône r. France **48** F3 47 30N 5 30E
São Paulo Brazil **84** E5 23 33S 46 39W
São Tomé i. Gulf of Guinea **66** C3 0 25N 6 35E
SÃO TOMÉ AND PRINCIPE 66 C3
Sapporo Japan **59** N3 43 05N 141 21E
Sarajevo Bosnia-Herzegovina **42** K4 43 52N 18 26E
Saratov Russia **56** E3 51 30N 45 55E
Sarawak admin. Malaysia **60** C3 1 00N 111 00E
Sardinia i. Italy **51** B4 40 00N 9 00E
Sarh Chad **66** D3 9 08N 18 22E
Sark i. Channel Islands British Isles **41** E2 49 26N 2 22W
Saskatchewan province Canada **78** J6 55 00N 105 00W
Saskatchewan r. Canada **78** J6 53 30N 105 00W
Saskatoon Canada **78** J6 52 10N 106 40W
Sassari Italy **51** B4 40 43N 8 34E
SAUDI ARABIA 61 E3
Saumur France **48** C3 47 16N 0 05W
Sava r. Europe **51** F6 45 45N 16 15E
Savannah USA **79** L4 32 04N 81 07W
Sawel mt. Northern Ireland **32** B2 54 49N 7 02W
Say'un Yemen Republic **61** E2 15 59N 48 44E
Scafell Pike mt. England **34** B3 54 27N 3 14W
Scalasaig Scotland **32** C4 56 04N 6 12W
Scalloway Scotland **29** D3 60 08N 1 17W
Scalpay i. Highland Scotland **30** D2 57 15N 6 00W
Scalpay i. Western Isles Scotland **30** C2 57 15N 6 00W
Scapa Scotland **29** C1 58 58N 2 59W
Scapa Flow sd. Scotland **29** B1 58 55N 3 00W
Scarba i. Scotland **32** D4 56 45N 5 50W
Scarborough England **35** E3 54 17N 0 24W
Scarp i. Scotland **30** B3 58 02N 7 08W
Schefferville Canada **78** N6 54 50N 67 00W
Schelde r. Netherlands **49** B4/C4 51 00N 4 00E
Schwerin Germany **49** F5 53 38N 11 25E
Scilly, Isles of England **40** A2 49 56N 6 20W
Scotland country United Kingdom **28** 56 10N 4 00W
Scottish Borders u.a. Scotland **33** F3/G3 55 30N 3 00W
Scunthorpe England **35** E2 53 35N 0 39W
Seascale England **34** B3 54 24N 3 29W
Seaton England **41** D3 50 43N 3 05W
Seattle USA **78** G5 47 35N 122 20W
Sedan France **48** F4 49 42N 4 57E
Seine r. France **48** D4 49 15N 1 15E
Sekondi Takoradi Ghana **66** A4 4 59N 1 43W
Selby England **35** D2 53 48N 1 04W
Selkirk Scotland **33** G3 55 33N 2 50W
Selsey Bill p. England **38** D1 50 43N 0 48W
Selvas geog. reg. South America **84** C7 7 00S 65 00W
Semarang Indonesia **60** C2 6 58S 110 29E
Semipalatinsk Kazakhstan **56** J3 50 26N 80 16E
Sendai Japan **59** N2 38 16N 140 52E
SENEGAL 66 A4
Senegal r. Senegal/Mauritania **66** A4 16 00N 15 00W
Sennar Sudan **61** D2 13 31N 33 38E
Sennen England **40** B3 50 03N 5 42W
Seoul South Korea **59** D3 37 32N 127 00E
Sept Îles r. Canada **79** N6 50 10N 66 00W
Seram i. Indonesia **60** D2 3 30S 129 30E
Seram Sea Indonesia **60** D2 2 30S 130 00E
SERBIA AND MONTENEGRO 42 L4
Seremban Malaysia **60** B3 2 43N 102 57E
Serov Russia **56** G3 59 42N 60 32E
Sétif Algeria **42** H3 36 11N 5 24E
Setúbal Portugal **50** A3 38 31N 8 54W
Sevastopol Ukraine **56** D2 44 36N 33 31E
Sevenoaks England **39** E2 51 16N 0 12E
Severn r. England **37** E1 51 30N 2 30W
Severnaya Zemlya is. Russia **57** K5 80 00N 95 00E
Severodvinsk Russia **43** P8 64 35N 39 50E
Seville Spain **50** C3 37 24N 5 59W
Seward USA **78** E7 60 05N 149 34W
SEYCHELLES 5
Sfax Tunisia **66** D6 34 45N 10 43E
Sgurr Mór mt. Scotland **30** D2 57 42N 5 03W
Shaftesbury England **41** E4 51 01N 2 12W
Shakhty Russia **43** Q5 47 43N 40 16E
Shanghai China **59** D3 31 06N 121 22E
Shanklin England **38** C1 50 38N 1 10W
Shannon r. Republic of Ireland **32** B2 52 45N 8 00W
Shapinsay i. Scotland **29** C2 59 03N 2 51W
Sheerness England **39** E2 51 27N 0 45E
Sheffield England **35** D2 53 23N 1 30W
Shenyang China **59** D4 41 50N 123 26E
Shepton Mallet England **41** E4 51 12N 2 33W
Sherborne England **41** E3 50 57N 2 31W
Sherbrooke Canada **79** M5 45 24N 71 54W
Sheringham England **39** F3 52 57N 1 12E
Shetland Islands u.a. Scotland **29** D3 60 00N 1 15W
Shiant Islands Scotland **30** C2 57 55N 6 20W
Shiel r. Scotland **30** D2 57 09N 5 25E
Shijiazhuang China **59** C3 38 04N 114 28E
Shikoku i. Japan **59** L3 33 40N 134 00E
Shilka r. Russia **57** M3 52 00N 118 00E
Shimonoseki Japan **59** L3 33 59N 130 58E
Shiraz Iran **61** F3 29 38N 52 34E
Shizuoka Japan **59** M1 34 59N 138 24E
Shreveport USA **79** J4 32 30N 93 46W
Shrewsbury England **37** E2 52 43N 2 45W
Shropshire co. England **37** E2 52 35N 2 25W
Shymkent Kazakhstan **56** G2 42 16N 69 05E
Šiauliai Lithuania **42** L7 55 51N 23 20E
Siberian Lowland Russia **56** H4 60 00N 75 00E
Sibu Malaysia **60** C3 2 19N 111 50E
Sichuan Basin see Sichuan Pendi
Sichuan Pendi (Sichuan Basin) China **59** B3 32 00N 107 00E
Sicily i. Italy **51** E2 37 00N 14 00E
Sidi-Bel-Abbès Algeria **42** F3 35 15N 0 39W
Sidlaw Hills Scotland **31** F1/G1 56 30N 3 10W
Sidmouth England **41** D3 50 41N 3 15W
Siegen Germany **49** E4 50 52N 8 02E
Siena Italy **51** C5 43 19N 11 19E
SIERRA LEONE 66 A3

Sierra Madre mts. Mexico **79** J3 26 00N 108 00W
Sierra Nevada mts. Spain **50** D3 37 00N 3 20W
Sierra Nevada mts. USA **79** H4 37 00N 119 00W
Simferopol Ukraine **56** D2 44 57N 34 05E
Simplon Pass Europe **49** E2 46 15N 8 03E
Simpson Desert Australia **72** D3 25 00S 137 00E
Sinai, Mount Egypt **61** D3 28 32N 33 59E
SINGAPORE 60 B3
Singkep i. Indonesia **60** B2 0 20S 104 30E
Sinuiju North Korea **59** D4 40 04N 124 25E
Siracusa Italy **51** E2 37 04N 15 19E
Sirte Libya **42** K2 31 13N 16 35E
Sivas Turkey **43** P3 39 44N 37 01E
Skegness England **35** F2 53 10N 0 21E
Skelmersdale England **34** C2 53 55N 2 48W
Skiddaw mt. England **34** B3 54 40N 3 08W
Skikda Algeria **42** H3 36 53N 6 54E
Skipton England **34** C2 53 58N 2 01W
Skopje FYROM **42** L4 42 00N 21 28E
Skye i. Scotland **30** C2 57 20N 6 15W
Sleaford England **35** E1 53 00N 0 24W
Sleat, Sound of Scotland **30** D2 57 00N 5 55W
Slieve Donard mt. Northern Ireland **32** D2 54 11N 5 55W
Slieve Snaght sum. Republic of Ireland **32** B3 55 12N 7 20W
Slough England **38** D2 51 31N 0 36W
Slough u.a. England **38** D2 51 31N 0 36W
SLOVAKIA 42 K5
SLOVENIA 51 E7
Smallwood Reservoir Canada **78** N6 54 10N 63 50W
Smolensk Russia **56** D3 54 49N 32 04E
Snaefell mt. Isle of Man British Isles **33** E2 54 16N 4 28W
Snake r. USA **79** H5 43 00N 114 00W
Snowdon mt. Wales **36** C2 53 04N 4 05W
Snowy Mountains Australia **72** E2 36 24S 148 28E
Soar r. England **37** F2 52 40N 1 10W
Soay i. Scotland **30** C2 57 08N 6 14W
Sochi Russia **43** P4 43 35N 39 46E
Socotra i. Yemen Republic **61** F2 12 05N 54 10E
Sofia Bulgaria **42** L4 42 40N 23 18E
Sokhumi Georgia **43** Q4 43 01N 41 01E
Sokoto Nigeria **66** C4 13 02N 5 15E
Solapur India **58** C2 17 43N 75 56E
Solihull England **37** F2 52 25N 1 45W
SOLOMON ISLANDS 72 F5
Solway Firth est. Scotland/England **34** B3 54 40N 3 40W
SOMALIA 67 E5
Somerset co. England **41** D4/E4 51 10N 3 00W
Somme r. France **48** E4 49 45N 2 30E
Songhua Hu res. China **59** D4 43 30N 126 51E
Songhua Jiang r. China **59** D4 46 30N 129 00E
Songkhla Thailand **60** B3 7 12N 100 35E
Song-koi r. China/Vietnam **59** B2 23 00N 102 00E
Sousse Tunisia **66** D6 35 50N 10 38E
Southampton England **38** C1 50 55N 1 25W
Southampton u.a. England **38** C1 50 54N 1 27W
Southampton Island Canada **78** L7 64 50N 85 00W
South Australia state Australia **72** D3 27 00S 135 00E
South Ayrshire u.a. Scotland **32/33** D2/E3 55 20N 4 40W
South Barrule hill Isle of Man British Isles **33** E2 54 09N 4 40W
South Benfleet England **39** E2 51 33N 0 34E
South Carolina state USA **79** L4 34 00N 81 00W
South China Sea Pacific Ocean **60** C4 15 00N 110 00E
South Dakota state USA **79** J5 45 00N 102 00W
South Downs hills England **38/39** D1/E 150 50N 0 45W
Southend Scotland **32** D3 55 20N 5 38W
Southend u.a. England **39** E2 51 33N 0 45E
Southend-on-Sea England **39** E2 51 33N 0 43E
Southern Alps mts. New Zealand **73** C2 43 07S 171 13E
Southern Ocean 86
Southern Uplands Scotland **33** E3/G3 55 30N 3 30W
South Esk r. Scotland **31** G1 56 40N 2 40W
South Georgia i. South Atlantic Ocean **85** F2 54 00S 36 00W
South Gloucestershire u.a. England **38** B2 51 30N 2 30W
South Island New Zealand **73** C2 42 30S 172 00E
SOUTH KOREA 59 D3
South Lanarkshire u.a. Scotland **33** E3/F3 55 35N 3 55W
South Molton England **40** D4 51 01N 3 50W
South Orkney Islands Southern Ocean **85** E1 60 00S 45 00W
South Pole Antarctica **86** 90 00S
Southport England **34** B2 53 39N 3 01W
South Ronaldsay i. Scotland **29** C1 58 47N 2 56W
South Shields England **35** D3 55 00N 1 25W
South Uist i. Scotland **30** B2 57 20N 7 15W
Southwest Cape New Zealand **73** B1 48 00S 168 00E
Southwold England **39** F3 52 20N 1 40E
South Yorkshire co. England **34/35** D2 53 30N 1 30W
Sovetskaya Gavan Russia **57** Q2 48 57N 140 16E
SPAIN 50 C3
Spalding England **35** E1 52 47N 0 10W
Spencer Gulf Australia **72** D2 34 16S 136 53E
Spennymoor tn. England **35** D3 54 42N 1 35W
Spey r. Scotland **31** F2 57 40N 3 10W
Spilsby England **35** F2 53 10N 0 05E
Spitsbergen i. Arctic Ocean **86** 79 00N 15 00E
Split Croatia **42** K4 43 31N 16 28E
Spokane USA **78** H5 47 40N 117 25W
Spurn Head England **35** F2 53 36N 0 07E
SRI LANKA 58 D1
Srinagar Jammu & Kashmir **58** C4 34 08N 74 50E
Stafford England **37** E2 52 48N 2 07W
Staffordshire co. England **37** E2/F2 52 50N 2 00W

Staines England **38** D2 51 26N 0 30W
Stamford England **35** E1 52 39N 0 29W
Stanley Falkland Islands **85** D2 51 45S 57 56W
Start Bay England **40** D3 50 16N 3 35W
Start Point England **40** D3 50 13N 3 38W
Stavanger Norway **42** H7 58 58N 5 45E
Stavropol Russia **43** Q5 45 03N 41 59E
Stevenage England **38** D2 51 55N 0 14W
Stewart Island New Zealand **73** B1 46 55S 167 55E
Stinchar r. Scotland **33** E3 55 10N 4 45W
Stirling Scotland **33** F4 56 07N 3 57W
Stirling u.a. Scotland **33** E4 56 10N 4 15W
Stockbridge England **38** C2 51 07N 1 29W
Stockholm Sweden **42** K7 59 20N 18 05E
Stockport England **34** C2 53 25N 2 10W
Stockton-on-Tees England **35** D3 54 34N 1 19W
Stockton-on-Tees u.a. England **35** D3 54 33N 1 24W
Stoke-on-Trent England **37** E3 53 00N 2 10W
Stoke-on-Trent u.a. England **37** E3 53 02N 2 10W
Stonehaven Scotland **31** G1 56 58N 2 13W
Stornoway Scotland **30** C3 58 12N 6 23W
Storr, The mt. Scotland **30** C2 57 30N 6 11W
Stour r. Dorset England **41** E3 50 50N 2 10W
Stour r. Suffolk/Essex England **39** E2/F2 51 55N 0 55E
Stourbridge England **37** E2 52 27N 2 09W
Stourport-on-Severn England **37** E2 52 21N 2 16W
Stowmarket England **39** F3 52 11N 1 00E
Stow-on-the-Wold England **37** F2 51 56N 1 44W
Strabane Northern Ireland **32** B2 54 49N 7 27W
Strabane district Northern Ireland **32** B2 54 45N 7 30W
Stralsund Germany **49** G6 54 18N 13 06E
Stranraer Scotland **32** D2 54 55N 5 02W
Strasbourg France **48** G4 48 35N 7 45E
Stratford-upon-Avon England **37** F2 52 12N 1 41W
Strathy Point Scotland **31** E3 58 35N 4 02W
Stroma i. Scotland **31** F3 58 40N 3 08W
Stromness Scotland **29** B1 58 57N 3 18W
Stronsay i. Scotland **29** C2 59 07N 2 37W
Stronsay Firth sd. Scotland **29** C2 59 00N 2 55W
Stroud England **38** B2 51 45N 2 12W
Strumble Head c. Wales **36** B2 52 02N 5 04W
Sturt Desert Australia **72** E3 27 00S 139 00E
Stuttgart Germany **49** E3 48 47N 9 12E
Sucre Bolivia **84** C6 19 05S 65 15W
SUDAN 61 C2
Sudbury Canada **79** L5 46 30N 81 01W
Sudbury England **39** E2 52 02N 0 44E
Suez Egypt **61** D3 29 59N 32 33E
Suez Canal Egypt **61** D4 31 30N 32 20E
Suffolk co. England **39** F3 52 10N 1 00E
Sula Islands Indonesia **60** D2 1 50S 124 50E
Sulawesi i. Indonesia **60** D2 2 00S 120 00E
Sullom Voe inlet Scotland **29** D3 60 27N 1 20W
Sulu Sea The Philippines/Malaysia **60** D3 8 00N 120 00E
Sumatra i. Indonesia **60** A3/B2 0 00 100 00E
Sumba i. Indonesia **60** C2 10 00S 120 00E
Sumbawa i. Indonesia **60** C2 8 00S 118 00E
Sumburgh Head Scotland **29** D2 59 51N 1 16W
Sunda Strait Indonesia **60** B2 5 50S 105 30E
Sunderland England **35** D3 54 55N 1 23W
Superior, Lake Canada/USA **79** L5 48 00N 88 00W
Surabaya Indonesia **60** C2 7 14S 112 45E
Surakarta Indonesia **60** C2 7 32S 110 50E
Surat India **58** C2 21 10N 72 54E
SURINAME 84 D8
Surrey co. England **38** D2 51 25N 0 14W
Sutton England **38** D2 51 22N 0 12W
Sutton Coldfield England **37** F2 52 34N 1 48W
Sutton in Ashfield England **37** F3 53 08N 1 15W
Suzhou China **59** D3 31 21N 120 40E
Sverdrup Islands Canada **78** K8 79 00N 96 00W
Swaffham England **39** E3 52 39N 0 41E
Swale r. England **34** D3 54 30N 2 00W
Swanage England **41** F3 50 37N 1 58W
Swansea Wales **36** D1 51 38N 3 57W
Swansea u.a. Wales **36** C1 51 40N 4 05W
SWAZILAND 67 D2
SWEDEN 42 J7/L9
Swilly r. Republic of Ireland **32** B3 55 00N 7 30W
Swindon England **38** C2 51 34N 1 47W
Swindon u.a. England **38** C2 51 32N 1 46W
SWITZERLAND 49 D2/E2
Sydney Australia **72** F2 33 52S 151 13E
Sydney Canada **79** N5 46 10N 60 10W
Syktyvkar Russia **43** S8 61 42N 50 45E
Symbister Scotland **29** D3 60 21N 1 02W
Syr Darya r. Asia **56** G2 45 00N 66 00E
SYRIA 61 D4
Szczecin Poland **42** J6 53 25N 14 32E
Szeged Hungary **42** L5 46 15N 20 09E

T

Tabora Tanzania **67** D4 5 01S 32 48E
Tabriz Iran **61** E4 38 05N 46 18E
Tabuk Saudi Arabia **61** D3 28 33N 36 36E
Tacoma USA **78** G5 47 16N 122 30W
Tadcaster England **35** D2 53 53N 1 16W
Taegu South Korea **59** D3 35 52N 128 36E
Taejon South Korea **59** D3 36 20N 127 26E
Tagus r. Spain/Portugal **50** B3 39 30N 7 00W
Tahat, Mount Algeria **66** C5 23 18N 5 33E
Tai Hu l. China **59** D3 31 15N 120 10E
Tain Scotland **31** E2 57 48N 4 04W
Taipei Taiwan **59** D2 25 03N 121 30E
TAIWAN 59 D2
Taiyuan China **59** C3 37 50N 112 30E
Ta'izz Yemen Republic **61** E2 13 35N 44 02E
TAJIKISTAN 56 H1
Takamatsu Japan **59** L1 34 20N 134 01E
Takazze r. Africa **61** D2 14 00N 38 00E
Talaud Islands Indonesia **60** D3 4 00N 127 00E
Talca Chile **85** B4 35 28S 71 40W

Talcahuano Chile **85** B4 36 40S 73 10W
Tamanrasset Algeria **66** C5 22 50N 5 28E
Tamar r. England **40** C3 50 35N 4 15W
Tambov Russia **43** Q6 52 44N 41 28E
Tampa USA **79** L3 27 58N 82 38W
Tampere Finland **42** L8 61 32N 23 45E
Tampico Mexico **79** K3 22 18N 97 52W
Tamworth England **37** F2 52 39N 1 40W
Tana, Lake Ethiopia **61** D2 12 00N 37 20E
Tanga Tanzania **67** D4 5 07S 39 05E
Tanganyika, Lake East Africa **67** C4 7 00S 30 00E
Tangier Morocco **66** B6 35 48N 5 45W
Tangshan China **59** C3 39 37N 118 05E
Tanimbar Islands Indonesia **60** E2 7 30S 131 30E
Tanjungkarang Indonesia **60** B2 5 28S 105 16E
TANZANIA 67 D4
Tapajós r. Brazil **84** D7 9 10S 56 20W
Taransay i. Scotland **30** B2 57 55N 7 00W
Táranto Italy **51** F4 40 28N 17 15E
Tarbat Ness c. Scotland **31** F2 57 50N 3 45W
Tarbert Argyll and Bute Scotland **32** D3 55 52N 5 26W
Tarbert Western Isles Scotland **30** C2 57 54N 6 49W
Tarbes France **48** D1 43 14N 0 05E
Tarbet Scotland **33** E4 56 12N 4 43W
Tarim He r. China **58** D4/5 40 00N 77 00E
Tarim He r. China **58** D4/5 40 00N 77 00E
Tarim Pendi (Tarim Basin) China **58** D4/5 39 00N 84 00E
Tarn r. France **48** E1 44 00N 2 15E
Tarragona Spain **50** F4 41 07N 1 15E
Tashkent Uzbekistan **56** G2 41 16N 69 13E
Tasman Bay New Zealand **73** C2 41 00S 173 14E
Tasmania state Australia **72** E1 42 00S 143 00E
Tasman Sea Pacific Ocean **7** 39 00S 155 00E
Taunton England **41** D4 51 01N 3 06W
Taupo New Zealand **73** D3 38 42S 176 05E
Taupo, Lake New Zealand **73** D3 38 55S 175 49E
Tauranga New Zealand **73** D3 37 41S 176 10E
Tavistock England **40** C3 50 33N 4 08W
Tavy r. England **40** C3 50 35N 4 05W
Taw r. England **40** C4/D3 51 00N 4 00W
Tay r. Scotland **31** F1 56 30N 3 30W
T'bilisi Georgia **56** E2 41 43N 44 48E
Te Anau, Lake New Zealand **73** B1 45 13S 167 45E
Tébessa Algeria **42** H3 35 21N 8 06E
Tees r. England **35** D3 54 30N 1 25W
Tegucigalpa Honduras **79** L2 14 05N 87 14W
Tehran Iran **61** F4 35 40N 51 26E
Teifi r. Wales **36** C2 52 03N 4 30W
Teign r. England **40** D3 50 40N 3 40W
Teignmouth England **40** D3 50 33N 3 30W
Tel Aviv-Yafo Israel **61** D4 32 05N 34 46E
Telford England **37** E2 52 42N 2 28W
Telford and Wrekin u.a. England **37** E2 52 45N 2 30W
Teme r. England **37** E2 52 20N 2 30W
Temuco Chile **85** B4 38 45S 72 40W
Tenby Wales **36** C1 51 41N 4 43W
Tenerife i. Canary Islands **66** A5 28 15N 16 35W
Tennessee r. USA **79** L4 36 00N 88 20W
Tennessee state USA **79** L4 35 00N 87 00W
Teramo Italy **51** D5 42 40N 13 43E
Teresina Brazil **84** E7 5 09S 42 46W
Terni Italy **51** D5 42 34N 12 39E
Terracina Italy **51** D4 41 17N 13 15E
Test r. England **38** C2 51 10N 1 30W
Tétouan Morocco **66** B6 35 34N 5 22W
Teviot r. Scotland **33** G3 55 30N 2 45W
Tewkesbury England **38** B2 51 59N 2 09W
Texas state USA **79** K4 31 00N 100 00W
THAILAND 60 B4
Thailand, Gulf of Asia **60** B4 10 50N 101 00E
Thame England **38** D2 51 45N 0 59W
Thame r. England **38** D2 51 45N 1 00W
Thames r. England **39** E2 51 30N 0 05E
Thar Desert India **58** C3 27 30N 72 00E
The Hague Netherlands **49** C5 52 05N 4 16E
The Needles rocks England **38** C1 50 39N 1 35W
The Peak mt. England **37** F3 53 24N 1 51W
The Solent sd. England **38** C1 50 45N 1 20W
Thessaloniki Greece **42** L4 40 38N 22 58E
Thetford England **39** E3 52 25N 0 45E
The Vale of Glamorgan u.a. Wales **36** D1 52 30N 3 30W
The Wrekin pk. England **37** E2 52 41N 2 32W
Thimphu Bhutan **58** D3 27 32N 89 43E
Thionville France **48** G4 49 22N 6 11E
Thirsk England **35** D3 54 14N 1 20W
Thornaby-on-Tees England **35** D3 54 34N 1 18W
Thornhill Scotland **33** F3 55 15N 3 46W
Three Kings Islands New Zealand **73** C4 34 10S 172 07E
Thunder Bay tn. Canada **79** L5 48 27N 89 12W
Thurrock England **39** E2 51 28N 0 20E
Thurrock u.a. England **39** E2 51 30N 0 23E
Thurso Scotland **31** F3 58 35N 3 32W
Thurso r. Scotland **31** F3 58 15N 3 35W
Tianjin China **59** C3 39 08N 117 12E
Tibetan Plateau Tibet **58** D4 33 00N 85 00E
Tibesti mts. Chad **66** D5 21 00N 17 00E
Tien Shan mts. China **58** C5/D5 41 00N 76 00E
Tierra del Fuego i. Chile/Argentina **85** C2 54 00S 67 30W
Tighnabruaich Scotland **32** D3 55 56N 5 14W
Tigris r. Iraq **61** E4 34 00N 43 00E
Tijuana Mexico **79** H4 32 29N 117 10W
Tilburg Netherlands **49** C4 51 34N 5 05E
Timaru New Zealand **73** C2 44 23S 171 15E
Timişoara Romania **42** L5 45 45N 21 15E
Timor i. Indonesia **60** D2 9 00S 125 00E
Timor Sea Australia **72** C4 12 30S 127 00E
Tiranë Albania **42** K4 41 20N 19 49E
Tiree i. Scotland **30** C1 56 30N 6 55W
Titicaca, Lake Peru/Bolivia **84** C6 16 00S 69 30W
Tiverton England **40** D3 50 55N 3 29W
Tivoli Italy **51** D4 41 58N 12 48E
Tizi Ouzou Algeria **50** H2 36 44N 4 05E
Toamasina Madagascar **67** E3 18 10S 49 23E

Tobago *i.* Trinidad and Tobago 84 C9 11 15N 60 40W
Tobermory Scotland 30 C1 56 37N 6 05W
Tobol *r.* Russia 56 G3 56 00N 66 00E
Tocantins *r.* Brazil 84 E7 8 40S 48 00W
TOGO 66 C3
Tokoroa New Zealand 73 D3 38 14S 175 52E
Tokyo Japan 59 M2 35 40N 139 45E
Toledo USA 79 L5 41 40N 83 35W
Toliara Madagascar 67 E2 23 20S 43 41E
Tol'yatti Russia 56 E3 53 32N 49 24E
Tombouctou Mali 66 B4 16 49N 2 59W
Tomsk Russia 56 J3 56 30N 85 05E
Tonbridge England 39 E2 51 12N 0 16E
Tone *r.* England 41 D4 50 44N 3 00W
Tone *r.* Japan 59 N2 36 00N 140 00E
TONGA 5
Toowoomba Australia 72 F3 27 34S 151 58E
Torbay England 40 D3 50 27N 3 30W
Torbay *u.a.* England 40 D3 50 26N 3 30W
Tordesillas Spain 50 C4 41 30N 5 00W
Torfaen *u.a.* Wales 36 D1 51 45N 3 03W
Toronto Canada 79 M5 43 42N 79 46W
Toros Dağları *mts.* Turkey 43 N3 36 00N 33 00E
Torpoint *tn.* England 40 C3 50 22N 4 11W
Torre del Greco Italy 51 E4 40 46N 14 22E
Torrens, Lake Australia 72 D2 30 58S 137 52E
Torreón Mexico 79 J3 25 34N 103 25W
Torres Strait Australia 72 E5 9 30S 142 00E
Torridge *r.* England 40 C3 50 55N 4 15W
Tortosa Spain 50 F4 40 49N 0 31E
Tory Island Republic of Ireland 32 A3 55 16N 8 14W
Tory Sound Republic of Ireland 32 A3 55 15N 8 05W
Totnes England 40 D3 50 25N 3 41W
Totton England 38 C1 50 56N 1 29W
Tottori Japan 59 M3 35 32N 134 12E
Toubkal, Mount Morocco 66 B6 31 03N 7 57W
Touggourt Algeria 66 C6 33 08N 6 04E
Toulon France 48 F1 43 07N 5 55E
Toulouse France 48 D1 43 33N 1 24E
Tours France 48 D3 47 23N 0 42E
Towcester England 38 D3 52 08N 1 00W
Townsville Australia 72 E4 19 17S 146 48E
Toyama Japan 59 M2 36 42N 137 14E
Trabzona Turkey 43 P4 41 00N 39 43E
Trent *r.* England 37 G3 53 30N 0 50W
Trento Italy 51 C7 46 04N 11 08E
Tresco *i.* England 40 A2 49 57N 6 20W
Treviso Italy 51 D6 45 40N 12 15E
Trevose Head England 40 B3 50 33N 5 01W
Trier Germany 49 D3 49 45N 6 39E
Trieste Italy 51 D6 45 39N 13 47E
Trincomalee Sri Lanka 58 D1 8 34N 81 13E
Trinidad *i.* Trinidad and Tobago 84 C9 11 00N 61 30W
TRINIDAD AND TOBAGO 79 N2
Tripoli Lebanon 61 D4 34 27N 35 50E
Tripoli Libya 66 D6 32 54N 13 11E
Tristan da Cunha *i.* Atlantic Ocean 5 37 15S 12 30W
Trivandrum India 58 C1 8 30N 76 57E
Tromsø Norway 42 K9 69 42N 19 00E
Trondheim Norway 42 J8 63 36N 10 23E
Troon Scotland 33 E3 55 32N 4 40W
Trowbridge England 38 B2 51 20N 2 13W
Troyes France 48 F4 48 18N 4 05E
Trujillo Peru 84 B7 8 06S 79 00W
Trujillo Spain 50 C3 39 28N 5 53W
Truro England 40 B3 50 16N 5 03W
Tsimlyansk Reservoir Russia 43 Q5 47 30N 43 00E
Tsugaru Channel Japan 59 N3 41 30N 140 30E
Tucson USA 79 H4 32 15N 110 57W
Tula Russia 56 D3 54 11N 37 38E
Tulsa USA 79 K4 36 07N 95 58W
Tunis Tunisia 66 D6 36 50N 10 13E
TUNISIA 66 C6
Turin Italy 51 A6 45 04N 7 40E
Turkana, Lake Ethiopia/Kenya 67 D5 4 00N 36 00E
TURKEY 43 N3
TURKMENISTAN 56 F1
Turks and Caicos Islands West Indies 79 M3 21 30N 72 00W
Turku Finland 42 L8 60 27N 22 15E
Turpan Depression China 58 D5/E5 42 40N 89 30E
Turriff Scotland 31 G2 57 32N 2 28W
TUVALU 5
Túy Spain 50 A5 42 03N 8 39W
Tver Russia 56 D3 56 49N 35 57E
Tweed *r.* Scotland/England 33 G3 55 45N 2 10W
Tynda Russia 57 N3 55 10N 124 35E
Tyndrum Scotland 31 E1 57 27N 4 44W
Tyne *r.* England 34 D3 54 58N 2 00W
Tyne and Wear *co.* England 35 D3 54 55N 1 20W
Tynemouth England 35 D4 55 01N 1 24W
Tyrrhenian Sea Europe 51 C4/D4 40 00N 12 00E
Tyumen Russia 56 G3 57 11N 65 29E
Tywi *r.* Wales 36 C1 51 45N 3 20W

U

Uberaba Brazil 84 E6 19 47S 47 57W
Uberlândia Brazil 84 E6 18 57S 48 17W
Ucayali *r.* Peru 84 B7 7 30S 75 00W
Udine Italy 51 D7 46 04N 13 14E
Uele *r.* Congo Dem. Rep. 67 C5 4 00N 26 00E
Ufa Russia 56 F3 54 45N 55 58E
UGANDA 67 D5
Ujung Pandang Indonesia 60 C2 5 09S 119 28E
Ukhta Russia 43 S8 63 33N 53 44E
UKRAINE 56 D2
Ulan Bator Mongolia 57 L2 47 54N 106 52E
Ulan-Ude Russia 57 L3 51 55N 107 40E
Ullapool Scotland 30 D2 57 54N 5 10W
Ullswater *l.* England 34 C3 54 35N 2 55W
Ullung Island South Korea 59 M2 37 30N 132 00E
Ulm Germany 49 F3 48 24N 10 00E
Ulva *i.* Scotland 30 C1 56 30N 6 15W

Ul'yanovsk Russia 56 E3 54 19N 48 22E
Umeå Sweden 42 L8 63 50N 20 15E
Unimak Island 78 C6 55 00N 164 00W
UNITED ARAB EMIRATES 61 F3
UNITED KINGDOM 42 F6
UNITED STATES OF AMERICA *(USA)* 79
Unst *i.* Scotland 29 E3 60 45N 0 55W
Upper Lough Erne *l.* Northern Ireland 32 B2 54 15N 7 30W
Uppsala Sweden 42 K7 59 55N 17 38E
Ural Mountains Russia 56 F3 57 00N 60 00E
Ural'sk Kazakhstan 56 F3 51 19N 51 20E
Ure *r.* England 35 D3 54 20N 2 00W
Urmia, Lake Iran 61 E4 37 20N 45 55E
URUGUAY 85 D4
Uruguay *r.* South America 85 D5 28 30S 55 30W
Ürümqi China 58 D5 43 43N 87 38E
Usk Wales 36 D1 51 59N 3 35W
Ustinov Russia 56 F3 56 49N 53 11E
Ust-Kut Russia 57 L3 56 48N 105 42E
Utah *state* USA 79 H4 39 00N 112 00W
Utrecht Netherlands 49 C5 52 05N 5 07E
Utsunomiya Japan 59 M2 36 33N 139 52E
Uttoxeter England 37 F2 52 54N 1 52W
UZBEKISTAN 56 G2

V

Vaal *r.* Republic of South Africa 67 C2 27 30S 25 00E
Vaasa Finland 42 L8 63 06N 21 36E
Vadodara India 58 C3 22 19N 73 14E
Vaduz Liechtenstein 49 F2 47 08N 9 32E
Valdepeñas Spain 50 D3 38 46N 3 24W
Valdivia Chile 85 B4 39 46S 73 15W
Valence France 48 F2 44 56N 4 54E
Valencia Spain 50 E3 39 29N 0 24W
Valencia Venezuela 84 C9 10 14N 67 59W
Valenciennes France 48 E5 50 22N 3 32E
Valladolid Spain 50 C4 41 39N 4 45W
Valletta Malta 51 E1 35 54N 14 32E
Valognes France 41 F2 49 31N 1 28W
Valparaíso Chile 85 B4 33 05S 71 40W
Vancouver Canada 78 G5 49 13N 123 06W
Vancouver Island Canada 78 G5 50 00N 124 33W
Vänern, Lake Sweden 42 J7 59 00N 13 30E
Van, Lake Turkey 43 Q3 38 33N 42 46E
Vannes France 48 C3 47 40N 2 44W
Vantaa Finland 42 L8 60 20N 24 50E
VANUATU 5
Varanasi India 58 D3 25 20N 83 00E
Varaždin Croatia 51 F7 46 18N 16 21E
Varna Bulgaria 42 M4 43 12N 27 57E
VENEZUELA 84 C8
Venice Italy 51 D6 45 26N 12 20E
Veracruz Mexico 79 K2 19 11N 96 10W
Verdun France 48 F4 49 10N 5 24E
Verkhoyansk Russia 57 P4 67 35N 133 25E
Vermont *state* USA 79 M5 44 00N 73 00W
Verona Italy 51 C6 45 26N 11 00E
Versailles France 48 E4 48 48N 2 07E
Vesuvius *mt.* Italy 51 E4 40 49N 14 26E
Veurne Belgium 39 G2 51 04N 2 40E
Vicenza Italy 51 C6 45 33N 11 32E
Vichy France 48 E3 46 07N 3 25E
Victoria Canada 78 G5 48 26N 123 20W
Victoria *r.* Australia 72 D4 17 00S 131 00E
Victoria *state* Australia 72 E2 47 00S 143 00E
Victoria Falls Zambia/Zimbabwe 67 C3 17 55S 25 51E
Victoria Island Canada 78 H8 70 45N 115 00W
Victoria, Lake East Africa 67 D4 2 00S 33 00E
Vienna Austria 42 K5 48 13N 16 22E
Vienne *r.* France 48 D3 46 30N 0 30E
Vientiane Laos 60 B4 17 59N 102 38E
Vierzon France 48 E3 47 14N 2 03E
VIETNAM 60 B4
Vigo Spain 50 A5 42 15N 8 44W
Vijayawada India 58 D2 16 34N 80 40E
Villach Austria 51 D7 46 37N 13 51E
Vilnius Lithuania 42 M6 54 40N 25 19E
Viña del Mar Chile 85 B4 33 02S 71 35W
Vinh Vietnam 60 B4 18 42N 105 41E
Vinnytsya Ukraine 56 C2 49 11N 28 30E
Vinson Massif *mts.* Antarctica 86 78 02S 22 00W
Vire *r.* France 41 F2 49 15N 1 05W
Virginia *state* USA 79 M4 38 00N 77 00W
Virgin Islands West Indies 79 N2 18 00N 64 30W
Visakhapatnam India 58 D2 17 42N 83 24E
Vitória Brazil 84 E5 20 20S 40 18W
Vitoria Spain 50 D5 42 51N 2 40W
Vitsyebsk Belarus 56 D3 55 10N 30 14E
Vladikavkaz Russia 43 Q4 43 02N 44 43E
Vladimir Russia 56 E3 56 08N 40 25E
Vladivostok Russia 57 P2 43 09N 131 53E
Volga *r.* Russia 56 E3 51 30N 48 00E
Volgograd Russia 56 E2 48 45N 44 30E
Vologda Russia 56 D3 59 10N 39 55E
Volta, Lake Ghana 66 B3 7 30N 0 30W
Vorkuta Russia 56 G4 67 27N 64 00E
Voronezh Russia 56 D3 51 40N 39 13E
Vosges *mts.* France 48 G4 48 10N 6 50E
Vyrnwy *r.* Wales 36 D2 52 45N 3 10W
Vyrnwy, Lake Wales 36 D2 52 47N 3 30W

W

Waal *r.* Netherlands 49 C4 51 50N 5 00E
Waddington, Mount Canada 78 G6 51 22N 125 14W
Wadebridge England 40 C3 50 32N 4 50W
Wadi Halfa Sudan 61 D3 21 55N 31 20E
Wad Medani Sudan 61 D2 14 24N 33 30E
Waiau *r.* New Zealand 73 B1 46 00S 167 30E
Waigeo *i.* Indonesia 60 E2 0 15S 130 45E
Waikato *r.* New Zealand 73 D3 37 30S 175 30E
Wairau *r.* New Zealand 73 C2 42 00S 173 00E
Waitaki *r.* New Zealand 73 C2 44 45S 170 30E
Wakatipu, Lake New Zealand 73 B1 45 06S 168 31E

Wakayama Japan 59 M1 34 12N 135 10E
Wakefield England 35 D2 53 42N 1 29W
Wakkanai Japan 59 N4 45 26N 141 43E
Walbury Hill England 38 C2 51 25N 1 35W
Wales *country* United Kingdom 36 D2 52 40N 3 30W
Wallasey England 34 B2 53 26N 3 03W
Walls Scotland 29 D3 60 14N 1 34W
Walsall England 37 F2 52 35N 1 58W
Waltham Forest England 38 D2 51 36N 0 00
Walton on the Naze England 39 F2 51 51N 1 16E
Walvis Bay *tn.* Namibia 67 B2 22 59S 14 31E
Wanaka, Lake New Zealand 73 B2 44 28S 169 09E
Wanganui New Zealand 73 D3 39 56S 175 03E
Wanganui *r.* New Zealand 73 D3 39 30S 175 00E
Wansbeck *r.* England 33 H3 55 10N 1 50W
Wantage England 38 C2 51 36N 1 25W
Ward Hill Scotland 29 B1 58 54N 3 20W
Ward's Stone *hill* England 34 C3 54 03N 2 38W
Wareham England 41 E3 50 41N 2 07W
Warley England 37 E2 52 30N 1 59W
Warminster England 38 B2 51 13N 2 12W
Warrenpoint *tn.* Northern Ireland 32 C2 54 06N 6 15W
Warrington England 34 C2 53 24N 2 37W
Warrington *u.a.* England 34 C2 53 23N 2 37W
Warsaw Poland 42 L6 52 15N 21 00E
Warwick England 37 F2 52 17N 1 34W
Warwickshire *co.* England 37 F2 52 15N 1 40W
Washington England 34 D3 54 54N 1 31W
Washington *state* USA 78 G5/H5 47 00N 120 00W
Washington D.C. USA 79 M4 38 55N 77 00W
Wash, The *b.* England 35 F1 52 55N 0 10E
Waterlooville England 38 C1 50 53N 1 02W
Water of Luce *r.* Scotland 32/33 E2 54 55N 4 55W
Watford England 38 D2 51 39N 0 24W
Wau Papua New Guinea 72 E5 7 22S 146 40E
Waveney *r.* England 39 F3 52 30N 1 30E
Weald, The *geog. reg.* England 39 E2 51 05N 0 25E
Wear *r.* England 34 C3 54 40N 2 09W
Weddell Sea Southern Ocean 86 71 00S 40 00W
Wei He *r.* China 59 B3 34 00N 106 00E
Welland *r.* England 38 D2 52 30N 0 40W
Wellingborough England 38 D3 52 19N 0 42W
Wellington England 41 D3 50 59N 3 15W
Wellington New Zealand 73 C2 41 17S 174 46E
Wells England 41 E4 51 13N 2 39W
Wells-next-the-Sea England 39 E3 52 58N 0 51E
Welshpool Wales 36 D2 52 40N 3 09W
Welwyn Garden City England 38 D2 51 48N 0 13W
Wenlock Edge *hills* England 37 E2 52 30N 2 45W
Wensum *r.* England 39 F3 52 45N 1 05E
Weser *r.* Germany 49 E5 52 30N 9 00E
West Berkshire *u.a.* England 38 C2 51 30N 1 17W
West Bromwich England 37 F2 52 31N 1 59W
Westbury England 38 B2 51 16N 2 11W
West Dunbartonshire *u.a.* Scotland 33 E3/4 56 03N 4 30W
Western Australia *state* Australia 72 25 00S 123 00E
Western Ghats *mts.* India 58 C2 15 30N 74 00E
Western Isles *u.a.* Scotland 30 B2/C3 57 30N 7 00W
WESTERN SAHARA 66 A5
West Friesian Islands Netherlands 49 C5 53 25N 5 15E
West Lothian *u.a.* Scotland 33 F3 55 50N 3 36W
West Midlands *co.* England 37 E2/F2 52 28N 1 40W
Weston-super-Mare England 38 B2 51 21N 2 59W
West Palm Beach *tn.* USA 79 L3 26 42N 80 05W
Westport New Zealand 73 C2 41 45S 171 36E
Westray *i.* Scotland 29 B2/C2 59 18N 3 00W
Westray Firth *sd.* Scotland 29 B2/C2 59 15N 3 00W
West Rift Valley Africa 67 D4 0 00 30 00E
West Sussex *co.* England 38 D1 51 00N 0 25W
West Virginia *state* USA 79 L4 39 00N 81 00W
West Yorkshire *co.* England 34/35 D2 53 50N 1 30W
Wetherby England 35 D2 53 56N 1 23W
Weymouth England 41 E3 50 37N 2 25W
Whakatane New Zealand 73 D3 37 58S 176 59E
Whalsay *i.* Scotland 29 E3 60 22N 0 59W
Whangarei New Zealand 73 D3 35 43S 174 19E
Wharfe *r.* England 34 C3 54 10N 2 05W
Whernside *sum.* England 34 C3 54 14N 2 23W
Whitby England 35 E3 54 29N 0 37W
Whitchurch England 37 E2 52 58N 2 41W
Whiteadder Water *r.* Scotland/England 33 G3 55 50N 2 10W
Whitehaven England 34 B3 54 33N 3 35W
Whitehorse Canada 78 F7 60 41N 135 08W
White Nile *r.* Sudan 61 D2 11 00N 32 00E
White Sea Russia 42 P9 66 00N 37 30E
Whithorn Scotland 33 E2 54 44N 4 25W
Whitley Bay *tn.* England 33 H3 55 03N 1 25W
Whitney, Mount USA 79 H4 36 35N 118 17W
Whyalla Australia 72 D2 33 02S 137 35E
Wichita USA 79 K4 37 43N 97 20W
Wick Scotland 31 F3 58 26N 3 06W
Wicklow Republic of Ireland 36 A2 52 59N 6 03W
Widnes England 34 C2 53 22N 2 44W
Wiesbaden Germany 49 E4 50 05N 8 15E
Wigan England 34 C2 53 33N 2 38W
Wigston England 37 F2 52 36N 1 05W
Wigton England 34 B3 54 49N 3 09W
Wigtown Scotland 33 E2 54 52N 4 26W
Wigtown Bay Scotland 33 E2 54 10N 4 20W
Wilhelmshaven Germany 49 E5 53 32N 8 07E
Wilkes Land *geog. reg.* Antarctica 86 68 00S 105 00E
Wiltshire *co.* England 38 B2/C2 51 30N 2 00W
Wimborne Minster England 41 F3 50 48N 1 59W
Wincanton England 41 E4 51 04N 2 25W
Winchester England 38 C2 51 04N 1 19W
Windermere England 34 C3 54 23N 2 54W
Windermere *l.* England 34 C3 54 20N 2 57W
Windhoek Namibia 67 B2 22 34S 17 06E
Windsor England 38 D2 51 29N 0 38W
Windsor & Maidenhead *u.a.* England 38 D2 51 30N 0 45W

Winnipeg Canada 79 K5 49 53N 97 10W
Winnipeg, Lake Canada 78 K6 53 00N 97 00W
Winsford England 34 C2 53 11N 2 31W
Winterthur Switzerland 49 E2 47 30N 8 45E
Wisbech England 39 E2 52 40N 0 10E
Wisconsin *state* USA 79 K5 45 00N 90 00W
Wishaw Scotland 33 F3 55 47N 3 56W
Witham England 39 E2 51 48N 0 38E
Witham *r.* England 35 E2 53 05N 0 10W
Witney England 38 C2 51 48N 1 29W
Wittenberg Germany 49 G5 51 53N 12 39E
Woking England 38 D2 51 20N 0 34W
Wokingham England 38 D2 51 25N 0 51W
Wokingham *u.a.* England 38 D2 51 24N 0 51W
Wolfsburg Germany 49 F5 52 27N 10 49E
Wollongong Australia 72 F2 34 25S 150 53E
Wolverhampton England 37 E2 52 36N 2 08W
Wonsan North Korea 59 D3 39 07N 127 26E
Woodbridge England 39 F3 52 06N 1 19E
Wooler England 33 G3 55 33N 2 01W
Worcester England 37 E2 52 11N 2 13W
Worcestershire *u.a.* England 37 E2 52 13N 2 13W
Workington England 34 B3 54 39N 3 33W
Worksop England 37 F3 53 18N 1 07W
Worms Head Wales 36 C1 51 34N 4 20W
Worthing England 38 D1 50 48N 0 23W
Wrangel *i.* Russia 57 U5 61 30N 180 00
Wrath, Cape Scotland 30 D3 58 37N 5 01W
Wrexham Wales 36 D3 53 03N 3 00W
Wrexham *u.a.* England 36 D3 53 00N 3 00W
Wrocław Poland 42 K6 51 05N 17 00E
Wuhan China 59 C3 30 35N 114 19E
Wuppertal Germany 49 D4 51 15N 7 10E
Würzburg Germany 49 E3 49 48N 9 57E
Wye *r.* England/Wales 37 E1 51 40N 2 40W
Wymondham England 39 F3 52 34N 1 07E
Wyndham Australia 72 C4 15 28S 128 06E
Wyoming *state* USA 79 J5 43 00N 108 00W
Wyre *r.* England 34 C2 53 50N 2 50W

X

Xi *r.* China 60 C5 23 00N 112 00E
Xi'an China 59 B3 34 16N 108 54E
Xi Jiang *r.* China 59 C2 22 00N 111 00E
Xingu *r.* Brazil 84 D7 5 30S 52 00W
Xining China 59 B3 36 35N 101 55E
Xuzhou China 59 C3 34 17N 117 18E

Y

Yakutsk Russia 57 N4 62 10N 129 50E
Yamagata Japan 59 N2 38 16N 140 16E
Yamoussoukro Côte d'Ivoire 66 B3 6 50N 5 20W
Yangon Myanmar 58 E2 16 45N 96 20E
Yangtze *r. see* Jinsha Jiang (upper course), Chang Jiang (lower course)
Yaoundé Cameroon 66 D3 3 51N 11 31E
Yarlung Zangbo Jiang *(Tsangpo)* *r.* China 58 D3 30 00N 90 00E
Yaroslavl Russia 56 D3 57 34N 39 52E
Yarrow Water *r.* Scotland 33 F3 55 35N 3 05W
Yazd Iran 61 F4 31 54N 54 22E
Yekaterinburg Russia 56 G3 56 52N 60 35E
Yell *i.* Shetland Islands Scotland 29 D3 60 35N 1 10W
Yellowknife Canada 78 H7 62 30N 114 29W
Yellow Sea China 59 D3 35 30N 122 30E
Yell Sound Scotland 29 D3 60 30N 1 15W
YEMEN REPUBLIC 61 E2
Yenisey *r.* Russia 57 J4 64 00N 88 00E
Yeo *r.* England 41 E4 51 00N 2 45W
Yeovil England 41 E3 50 57N 2 39W
Yerevan Armenia 43 Q4 40 10N 44 31E
Yes Tor *sum.* England 40 C3 50 42N 4 00W
Yogyakarta Indonesia 60 C2 7 48S 110 24E
Yokkaichi Japan 59 M1 34 58N 136 38E
Yokohama Japan 59 M2 35 27N 139 38E
York England 35 D2 53 58N 1 05W
York *u.a.* England 35 D2 53 58N 1 05W
York, Cape Australia 72 E4 10 41S 142 32E
Yorkshire Wolds *hills* England 35 E3 54 00N 0 45W
York, Vale of England 35 D3 54 10N 1 20W
Ypres *see* Ipres
Yucatan Channel Caribbean Sea 79 L3 21 00N 85 00W
Yucatan Peninsula Mexico 79 L2/3 19 00N 89 00W
Yukon *r.* Canada/USA 78 D7 65 00N 155 00W
Yukon Territory Canada 78 F7 62 00N 135 00W

Z

Zagreb Croatia 42 K5 45 48N 15 58E
Zagros Mountains Iran 61 E4/F3 32 45N 48 50E
Zahedan Iran 61 G3 29 32N 60 54E
Zahlé Lebanon 43 P2 33 50N 35 55E
Zambezi *r.* Southern Africa 67 D3 17 00S 34 00E
ZAMBIA 67 C3
Zamboanga The Philippines 60 D3 6 55N 122 05E
Zanzibar Tanzania 67 D4 6 10S 39 12E
Zaporizhzhya Ukraine 43 P5 47 50N 35 10E
Zaragoza Spain 50 E4 41 39N 0 54W
Zaria Nigeria 66 C4 11 01N 7 44E
Zarqa Jordan 43 P2 32 04N 36 05E
Zaysan, Lake Kazakhstan 56 J2 48 00N 84 00E
Zealand *i.* Denmark 49 F6 55 00N 12 00E
Zeebrugge Belgium 49 B4 51 20N 3 13E
Zhambyl Kazakhstan 56 H2 42 50N 71 25E
Zhangjiakou China 59 C4 40 51N 114 59E
Zhangye China 59 A4 100.30E
Zhanjiang China 59 C2 21 10N 110 20E
Zhengzhou China 59 C3 34 45N 113 38E
Zhytomyr Ukraine 42 M6 50 18N 28 40E
Zibo China 59 C3 36 51N 118 01E
Ziel, Mount Australia 72 D3 23 30S 132 24E
ZIMBABWE 67 C3
Zinder Niger 66 C4 13 46N 8 58E
Zonguldak Turkey 43 N4 41 26N 31 47E
Zürich Switzerland 49 E2 47 23N 8 33E
Zwickau Germany 49 G4 50 43N 12 30E

World Flags

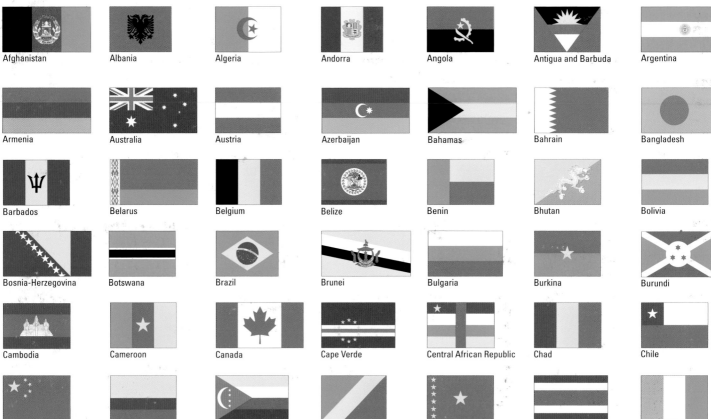

Afghanistan	Albania	Algeria	Andorra	Angola	Antigua and Barbuda	Argentina
Armenia	Australia	Austria	Azerbaijan	Bahamas	Bahrain	Bangladesh
Barbados	Belarus	Belgium	Belize	Benin	Bhutan	Bolivia
Bosnia-Herzegovina	Botswana	Brazil	Brunei	Bulgaria	Burkina	Burundi
Cambodia	Cameroon	Canada	Cape Verde	Central African Republic	Chad	Chile
China	Colombia	Comoros	Congo	Congo, Dem. Rep.	Costa Rica	Côte d'Ivoire
Croatia	Cuba	Cyprus	Czech Republic	Denmark	Djibouti	Dominica
Dominican Republic	East Timor	Ecuador	Egypt	El Salvador	Equatorial Guinea	Eritrea
Estonia	Ethiopia	Fiji	Finland	France	French Guiana	Gabon
Gambia	Georgia	Germany	Ghana	Greece	Greenland	Grenada
Guatemala	Guinea	Guinea-Bissau	Guyana	Haiti	Honduras	Hungary
Iceland	India	Indonesia	Iran	Iraq	Ireland	Israel
Italy	Jamaica	Japan	Jordan	Kazakhstan	Kenya	Kiribati
Kuwait	Kyrgyzstan	Laos	Latvia	Lebanon	Lesotho	Liberia

        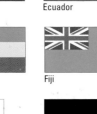            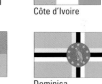   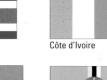